Calculus
FOR
DUMMIES
A Wiley Brand

2nd Edition

by Mark Ryan
Founder of The Math Center

FOR
DUMMIES
A Wiley Brand

Calculus For Dummies®, 2nd Edition

Published by: **John Wiley & Sons, Inc.,** 111 River Street, Hoboken, NJ 07030-5774, www.wiley.com

Copyright © 2014 by John Wiley & Sons, Inc., Hoboken, New Jersey

Published simultaneously in Canada

For general information on our other products and services, please contact our Customer Care Department within the U.S. at 877-762-2974, outside the U.S. at 317-572-3993, or fax 317-572-4002. For technical support, please visit www.wiley.com/techsupport.

Wiley publishes in a variety of print and electronic formats and by print-on-demand. Some material included with standard print versions of this book may not be included in e-books or in print-on-demand. If this book refers to media such as a CD or DVD that is not included in the version you purchased, you may download this material at http://booksupport.wiley.com. For more information about Wiley products, visit www.wiley.com.

Library of Congress Control Number: 2013958398

ISBN 978-1-118-79129-5 (pbk); ISBN 978-1-118-79108-0 (ePDF); ISBN 978-1-118-79133-2 (ePub)

Manufactured in the United States of America

10 9 8 7 6 5 4 3 2 1

Contents at a Glance

Table of Contents

Introduction

● ●

*T*he mere thought of having to take a required calculus course is enough to make legions of students break out in a cold sweat. Others who have no intention of ever studying the subject have this notion that calculus is impossibly difficult unless you happen to be a direct descendant of Einstein.

Well, I'm here to tell you that you *can* master calculus. It's not nearly as tough as its mystique would lead you to think. Much of calculus is really just very advanced algebra, geometry, and trig. It builds upon and is a logical extension of those subjects. If you can do algebra, geometry, and trig, you can do calculus.

But why should you bother — apart from being required to take a course? Why climb Mt. Everest? Why listen to Beethoven's Ninth Symphony? Why visit the Louvre to see the Mona Lisa? Why watch *South Park?* Like these endeavors, doing calculus can be its own reward. There are many who say that calculus is one of the crowning achievements in all of intellectual history. As such, it's worth the effort. Read this jargon-free book, get a handle on calculus, and join the happy few who can proudly say, "Calculus? Oh, sure, I know calculus. It's no big deal."

About This Book

Calculus For Dummies, 2nd Edition is intended for three groups of readers: students taking their first calculus course, students who need to brush up on their calculus to prepare for other studies, and adults of all ages who'd like a good introduction to the subject.

If you're enrolled in a calculus course and you find your textbook less than crystal clear, this is the book for you. It covers the most important topics in the first year of calculus: differentiation, integration, and infinite series.

If you've had elementary calculus, but it's been a couple of years and you want to review the concepts to prepare for, say, some graduate program, *Calculus For Dummies,* 2nd Edition will give you a thorough, no-nonsense refresher course.

Non-student readers will find the book's exposition clear and accessible. *Calculus For Dummies,* 2nd Edition takes calculus out of the ivory tower and brings it down to earth.

This is a user-friendly math book. Whenever possible, I explain the calculus concepts by showing you connections between the calculus ideas and easier ideas from algebra and geometry. I then show you how the calculus concepts work in concrete examples. Only later do I give you the fancy calculus formulas. All explanations are in plain English, not math-speak.

The following conventions keep the text consistent and oh-so-easy to follow:

- ✔ Variables are in *italics*.
- ✔ Calculus terms are italicized and defined when they first appear in the text.
- ✔ In the step-by-step problem-solving methods, the general action you need to take is in bold, followed by the specifics of the particular problem.

It can be a great aid to true understanding of calculus — or any math topic for that matter — to focus on the *why* in addition to the *how-to*. With this in mind, I've put a lot of effort into explaining the underlying logic of many of the ideas in this book. If you want to give your study of calculus a solid foundation, you should read these explanations. But if you're really in a hurry, you can cut to the chase and read only the important introductory stuff, the example problems, the step-by-step solutions, and all the rules and definitions next to the icons. You can then read the remaining exposition only if you feel the need.

I find the sidebars interesting and entertaining. (What do you expect? I wrote them!) But you can skip them without missing any essential calculus. No, you won't be tested on this stuff.

Minor note: Within this book, you may note that some web addresses break across two lines of text. If you're reading this book in print and want to visit one of these web pages, simply key in the web address exactly as it's noted in the text, as though the line break doesn't exist. If you're reading this as an e-book, you've got it easy — just click the web address to be taken directly to the web page.

Foolish Assumptions

Call me crazy, but I assume . . .

- ✔ You know at least the basics of algebra, geometry, and trig.

  If you're rusty, Part II (and the online Cheat Sheet) contains a good review of these pre-calculus topics. Actually, if you're not currently taking a calculus course, and you're reading this book just to satisfy a general curiosity about calculus, you can get a good conceptual picture of the subject without the nitty-gritty details of algebra, geometry, and trig. But you won't, in that case, be able to follow all the problem solutions. In short, without the pre-calculus stuff, you can see the calculus *forest*, but not the *trees*. If you're enrolled in a calculus course, you've got no choice — you've got to know the trees as well as the forest.

↙ You're willing to do some w_ _ _ .

No, not the dreaded *w*-word! Yes, that's w-o-r-k, *work*. I've tried to make this material as accessible as possible, but it is calculus after all. You can't learn calculus by just listening to a tape in your car or taking a pill — not yet anyway.

Is that too much to ask?

Icons Used in This Book

Keep your eyes on the icons:

Next to this icon are the essential calculus rules, definitions, and formulas you should definitely know.

These are things you need to know from algebra, geometry, or trig, or things you should recall from earlier in the book.

The bull's-eye icon appears next to things that will make your life easier. Take note.

This icon highlights common calculus mistakes. Take heed.

In contrast to the Critical Calculus Concepts, you generally don't need to memorize the fancy-pants formulas next to this icon unless your calc teacher insists.

Beyond the Book

There's some great supplementary calculus material online that you might want to check out:

- ✔ On the online Cheat Sheet, located at www.dummies.com/cheatsheet/ calculus, you'll find a nice list of important formulas, theorems, definitions, and so on from algebra, geometry, trigonometry, and calculus. This is a great place to go if you forget a formula.

- ✔ At www.dummies.com/extras/calculus, there are articles on some calculus topics that many calculus courses skip. For example, the online article, "Finding Volume with the Matryoshka Doll Method (a.k.a. the Cylindrical Shell Method)" covers one of the methods for computing volume that used to be part of the standard calculus curriculum, but which is now often omitted. You'll also find other interesting, off-the-beaten-path calculus articles. Check them out if you just can't get enough calculus.

Where to Go from Here

Why, Chapter 1, of course, if you want to start at the beginning. If you already have some background in calculus or just need a refresher course in one area or another, then feel free to skip around. Use the table of contents and index to find what you're looking for. If all goes well, in a half a year or so, you'll be able to check calculus off your list:

__ Run a marathon

__ Go skydiving

__ Write a book

✔ Learn calculus

__ Swim the English Channel

__ Cure cancer

__ Write a symphony

__ Pull an inverted 720° at the X-Games

For the rest of your list, you're on your own.

Part I

An Overview of Calculus

getting started with

Calculus

In this part . . .

- A brief and straightforward explanation of just what calculus is. Hint: it's got a lot to do with curves and with things that are constantly changing.

- Examples of where you might see calculus at work in the real world: curving cables, curving domes, and the curving path of a spacecraft.

- The first of the two big ideas in calculus: *differentiation,* which means finding a derivative. A derivative is basically just the fancy calculus version of a slope; and it's a simple rate — a this per that.

- The second big calculus idea: *integration.* It's the fancy calculus version of adding up small parts of something to get the total.

- An honest-to-goodness explanation of why calculus works: In short, it's because when you zoom in on curves (infinitely far), they become straight.

Chapter 1

What Is Calculus?

> *"My best day in Calc 101 at Southern Cal was the day I had to cut class to get a root canal."*
>
> — Mary Johnson

> *"I keep having this recurring dream where my calculus professor is coming after me with an axe."*
>
> — Tom Franklin, Colorado College sophomore

> *"Calculus is fun, and it's so easy. I don't get what all the fuss is about."*
>
> — Sam Einstein, Albert's great grandson

*I*n this chapter, I answer the question "What is calculus?" in plain English, and I give you real-world examples of how calculus is used. After reading this and the following two short chapters, you *will* understand what calculus is all about. But here's a twist: Why don't you start out on the *wrong* foot by briefly checking out what calculus is *not?*

What Calculus Is Not

No sense delaying the inevitable. Ready for your first calculus test? Circle True or False.

> True or False: Unless you actually enjoy wearing a pocket protector, you've got no business taking calculus.

> True or False: Studying calculus is hazardous to your health.

> True or False: Calculus is totally irrelevant.

False, false, false! There's this mystique about calculus that it's this ridiculously difficult, incredibly arcane subject that no one in their right mind would sign up for unless it was a required course.

Don't buy into this misconception. Sure, calculus is difficult — I'm not going to lie to you — but it's manageable, doable. You made it through algebra, geometry, and trigonometry. Well, calculus just picks up where they leave off — it's simply the next step in a logical progression.

And calculus is not a dead language like Latin, spoken only by academics. It's the language of engineers, scientists, and economists. Okay, so it's a couple steps removed from your everyday life and unlikely to come up at a cocktail party. But the work of those engineers, scientists, and economists has a huge impact on your day-to-day life — from your microwave oven, cell phone, TV, and car to the medicines you take, the workings of the economy, and our national defense. At this very moment, something within your reach or within your view has been impacted by calculus.

So What Is Calculus Already?

Calculus is basically just very advanced algebra and geometry. In one sense, it's not even a new subject — it takes the ordinary rules of algebra and geometry and tweaks them so that they can be used on more complicated problems. (The rub, of course, is that darn *other* sense in which it *is* a new and more difficult subject.)

Look at Figure 1-1. On the left is a man pushing a crate up a straight incline. On the right, the man is pushing the same crate up a curving incline. The problem, in both cases, is to determine the amount of energy required to push the crate to the top. You can do the problem on the left with regular math. For the one on the right, you need calculus (assuming you don't know the physics shortcuts).

Figure 1-1:
The difference between regular math and calculus: In a word, it's the *curve*.

Regular math problem   Calculus problem

For the straight incline, the man pushes with an *unchanging* force, and the crate goes up the incline at an *unchanging* speed. With some simple physics formulas and regular math (including algebra and trig), you can compute how many calories of energy are required to push the crate up the incline. Note that the amount of energy expended each second remains the same.

For the curving incline, on the other hand, things are constantly changing. The steepness of the incline is *changing* — and not just in increments like it's one steepness for the first 3 feet then a different steepness for the next 3 feet. It's *constantly changing*. And the man pushes with a *constantly changing* force — the steeper the incline, the harder the push. As a result, the amount of energy expended is also changing, not every second or every thousandth of a second, but *constantly changing* from one moment to the next. That's what makes it a calculus problem. By this time, it should come as no surprise to you that calculus is described as "the mathematics of change." Calculus takes the regular rules of math and applies them to fluid, evolving problems.

For the curving incline problem, the physics formulas remain the same, and the algebra and trig you use stay the same. The difference is that — in contrast to the straight incline problem, which you can sort of do in a single shot — you've got to break up the curving incline problem into small chunks and do each chunk separately. Figure 1-2 shows a small portion of the curving incline blown up to several times its size.

Figure 1-2:
Zooming in on the curve — voilà, it's straight (almost).

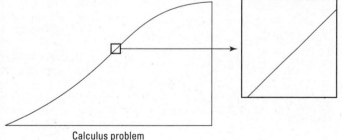

Calculus problem

When you zoom in far enough, the small length of the curving incline becomes practically straight. Then, because it's straight, you can solve that small chunk just like the straight incline problem. Each small chunk can be solved the same way, and then you just add up all the chunks.

That's calculus in a nutshell. It takes a problem that can't be done with regular math because things are constantly changing — the changing quantities show up on a graph as curves — it zooms in on the curve till it becomes straight, and then it finishes off the problem with regular math.

What makes the invention of calculus such a fantastic achievement is that it does what seems impossible: it zooms in *infinitely*. As a matter of fact, everything in calculus involves infinity in one way or another, because if something is constantly changing, it's changing infinitely often from each infinitesimal moment to the next.

Real-World Examples of Calculus

So, with regular math you can do the straight incline problem; with calculus you can do the curving incline problem. Here are some more examples.

With regular math you can determine the length of a buried cable that runs diagonally from one corner of a park to the other (remember the Pythagorean theorem?). With calculus you can determine the length of a cable hung between two towers that has the shape of a *catenary* (which is different, by the way, from a simple circular arc or a parabola). Knowing the exact length is of obvious importance to a power company planning hundreds of miles of new electric cable. See Figure 1-3.

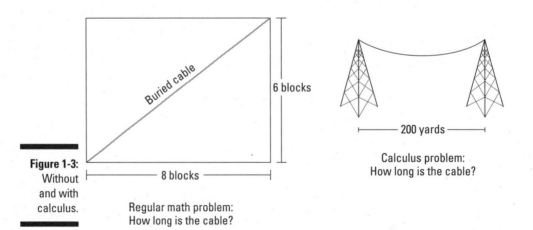

Figure 1-3: Without and with calculus.

6 blocks

8 blocks

Regular math problem: How long is the cable?

200 yards

Calculus problem: How long is the cable?

You can calculate the area of the flat roof of a home with ordinary geometry. With calculus you can compute the area of a complicated, nonspherical shape like the dome of the Minneapolis Metrodome. Architects designing such a building need to know the dome's area to determine the cost of materials and to figure the weight of the dome (with and without snow on it). The weight, of course, is needed for planning the strength of the supporting structure. Check out Figure 1-4.

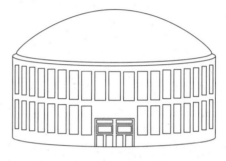

Figure 1-4: Sans and avec calculus.

Regular math problem: What's the roof's area?

Calculus problem: What's the dome's area?

With regular math and some simple physics, you can calculate how much a quarterback must lead his receiver to complete a pass. (I'm assuming here that the receiver runs in a *straight* line and at a *constant* speed.) But when NASA, in 1975, calculated the necessary "lead" for aiming the Viking I at Mars, it needed calculus because both the Earth and Mars travel on *elliptical* orbits (of different shapes) and the speeds of both are *constantly changing* — not to mention the fact that on its way to Mars, the spacecraft is affected by the different and *constantly changing* gravitational pulls of the Earth, the moon, Mars, and the sun. See Figure 1-5.

You see many real-world applications of calculus throughout this book. The differentiation problems in Part IV all involve the steepness of a curve — like the steepness of the curving incline in Figure 1-1. In Part V, you do integration problems like the cable-length problem shown back in Figure 1-3. These problems involve breaking up something into little sections, calculating each section, and then adding up the sections to get the total. More about that in Chapter 2.

Regular math problem:
What's the proper lead for
hitting the receiver?

Calculus problem:
What's the proper "lead" for
"hitting" Mars?

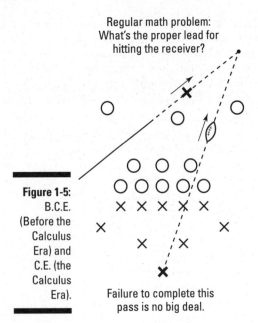

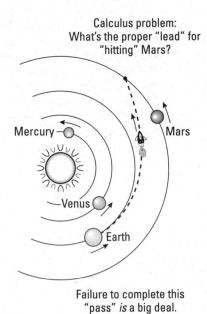

Figure 1-5:
B.C.E.
(Before the
Calculus
Era) and
C.E. (the
Calculus
Era).

Failure to complete this
pass is no big deal.

Failure to complete this
"pass" *is* a big deal.

Chapter 2

The Two Big Ideas of Calculus: Differentiation and Integration — plus Infinite Series

• •

In This Chapter

▶ Delving into the derivative: It's a rate and a slope

▶ Investigating the integral — addition for experts

▶ Infinite series: Achilles versus the tortoise — place your bets

• •

*T*his book covers the two main topics in calculus — differentiation and integration — as well as a third topic, infinite series. All three topics touch the earth and the heavens because all are built upon the rules of ordinary algebra and geometry and all involve the idea of infinity.

Defining Differentiation

Differentiation is the process of finding the *derivative* of a curve. And the word "derivative" is just the fancy calculus term for the curve's slope or steepness. And because the slope of a curve is equivalent to a simple rate (like *miles per hour* or *profit per item*), the derivative is a rate as well as a slope.

The derivative is a slope

In algebra, you learned about the slope of a line — it's equal to the ratio of the *rise* to the *run*. In other words, $Slope = \dfrac{rise}{run}$. See Figure 2-1. Let me guess: A sudden rush of algebra nostalgia is flooding over you.

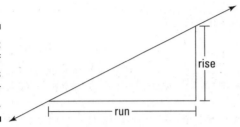

In Figure 2-1, the *rise* is half as long as the *run*, so the line has a slope of 1/2.

On a curve, the slope is constantly *changing,* so you need calculus to determine its slope. See Figure 2-2.

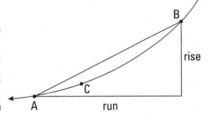

Just like the line in Figure 2-1, the straight line between A and B in Figure 2-2 has a slope of 1/2. And the slope of this line is the same at every point between A and B. But you can see that, unlike the line, the steepness of the curve is changing between A and B. At A, the curve is less steep than the line, and at B, the curve is steeper than the line. What do you do if you want the exact slope at, say, point C? Can you guess? Time's up. Answer: You zoom in. See Figure 2-3.

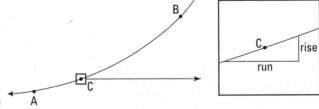

When you zoom in far enough — really far, actually *infinitely* far — the little piece of the curve becomes straight, and you can figure the slope the old-fashioned way. That's how differentiation works.

The derivative is a rate

Because the derivative of a curve is the slope — which equals $\frac{rise}{run}$ or *rise per run* — the derivative is also a rate, a *this per that* like *miles per hour* or *gallons per minute* (the name of the particular rate simply depends on the units used on the *x*- and *y*-axes). The two graphs in Figure 2-4 show a relationship between distance and time — they could represent a trip in your car.

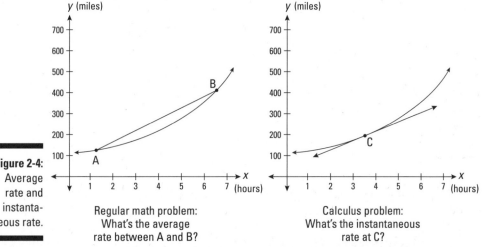

Figure 2-4: Average rate and instantaneous rate.

Regular math problem: What's the average rate between A and B?

Calculus problem: What's the instantaneous rate at C?

A regular algebra problem is shown on the left in Figure 2-4. If you know the *x*- and *y*-coordinates of points A and B, you can use the slope formula $\left(Slope = \frac{rise}{run} = \frac{y_2 - y_1}{x_2 - x_1} \right)$ to calculate the slope between A and B, and, in this problem, that slope gives you the *average* rate in *miles per hour* for the interval from A to B.

For the problem on the right, on the other hand, you need calculus. (You can't use the slope formula because you've only got one point.) Using the derivative of the curve, you can determine the *exact* slope or steepness at point C. Just to the left of C on the curve, the slope is slightly lower, and just to the right of C on the curve, the slope is slightly higher. But precisely at C, for a single infinitesimal moment, you get a slope that's different from the neighboring slopes. The slope for this single infinitesimal point on the curve gives you the *instantaneous* rate in *miles per hour* at point C.

Investigating Integration

Integration is the second big idea in calculus, and it's basically just fancy addition. Integration is the process of cutting up an area into tiny sections, figuring the areas of the small sections, and then adding up the little bits of area to get the whole area. Figure 2-5 shows two area problems — one that you can do with geometry and one where you need calculus.

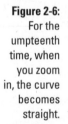

Figure 2-5: If you can't determine the area on the left, hang up your calculator.

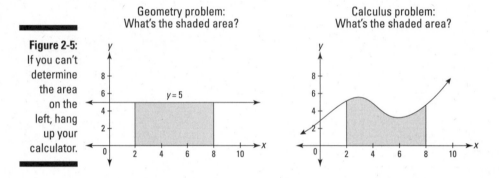

Geometry problem: What's the shaded area?

Calculus problem: What's the shaded area?

The shaded area on the left is a simple rectangle, so its area, of course, equals length times width. But you can't figure the area on the right with regular geometry because there's no area formula for this funny shape. So what do you do? Why, zoom in, of course. Figure 2-6 shows the top portion of a narrow strip of the weird shape blown up to several times its size.

Figure 2-6: For the umpteenth time, when you zoom in, the curve becomes straight.

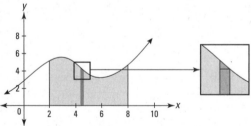

When you zoom in as shown in Figure 2-6, the curve becomes practically straight, and the further you zoom in, the straighter it gets. After zooming in, you get the shape on the right in Figure 2-6, which is practically an ordinary trapezoid (its top is still slightly curved). Well, with the magic of integration, you zoom in *infinitely* close (sort of — you can't really get infinitely close, right?). At that point, the shape is exactly an ordinary trapezoid — or, if you want to get really basic, it's a triangle sitting on top of a rectangle. Because you can compute the areas of rectangles, triangles, and trapezoids with ordinary geometry, you can get the area of this and all the other thin strips and then add up all these areas to get the total area. That's integration.

Figure 2-7 has two graphs of a city's electrical energy consumption on a typical summer day. The horizontal axes show the number of hours after midnight, and the vertical axes show the amount of power (in kilowatts) used by the city at different times during the day.

Figure 2-7: Total kilowatt-hours of energy used by a city during a single day.

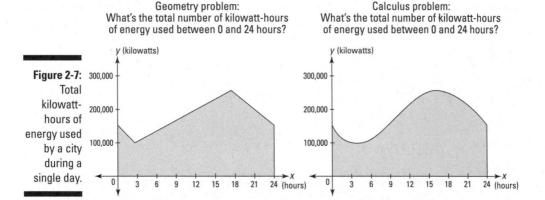

Geometry problem:
What's the total number of kilowatt-hours of energy used between 0 and 24 hours?

Calculus problem:
What's the total number of kilowatt-hours of energy used between 0 and 24 hours?

The crooked line on the left and the curve on the right show how the number of kilowatts of power depends on the time of day. In both cases, the shaded area gives the number of kilowatt-hours of energy consumed during a typical 24-hour period. The shaded area in the oversimplified and unrealistic problem on the left can be calculated with regular geometry. But the true relationship between the amount of power used and the time of day is more complicated than a crooked straight line. In a realistic energy-consumption problem, you'd get something like the graph on the right. Because of its weird curve, you need calculus to determine the shaded area. In the real world, the relationship between different variables is rarely as simple as a straight-line graph. That's what makes calculus so useful.

Sorting Out Infinite Series

Infinite series deal with the adding up of an infinite number of numbers. Don't try this on your calculator unless you've got a lot of extra time on your hands. Here's a simple example. The following sequence of numbers is generated by a simple doubling process — each term is twice the one before it:

1, 2, 4, 8, 16, 32, 64, 128, . . .

The infinite *series* associated with this *sequence* of numbers is just the sum of the numbers:

$$1 + 2 + 4 + 8 + 16 + 32 + 64 + 128 + \ldots$$

Divergent series

The preceding series of doubling numbers is *divergent* because if you continue the addition indefinitely, the sum will grow bigger and bigger without limit. And if you could add up "all" the numbers in this series — that's all *infinitely many* of them — the sum would be infinity. *Divergent* usually means — there are exceptions — that the series adds up to infinity.

Divergent series are rather uninteresting because they do what you expect. You keep adding more numbers, so the sum keeps growing, and if you continue this forever, the sum grows to infinity. Big surprise.

Convergent series

Convergent series are much more interesting. With a convergent series, you also keep adding more numbers, the sum keeps growing, but even though you add numbers forever and the sum grows forever, the sum of all the infinitely many terms is a *finite* number. This surprising result brings me to Zeno's famous paradox of Achilles and the tortoise. (That's Zeno of Elea, of course, from the 5th century B.C.)

Achilles is racing a tortoise — some gutsy warrior, eh? Our generous hero gives the tortoise a 100-yard head start. Achilles runs at 20 mph; the tortoise "runs" at 2 mph. Zeno used the following argument to "prove" that Achilles will never catch or pass the tortoise. If you're persuaded by this "proof," by the way, you've really got to get out more.

Imagine that you're a journalist covering the race for *Spartan Sports Weekly*, and you're taking a series of photos for your article. Figure 2-8 shows the situation at the start of the race and your first two photos.

You take your first photo the instant Achilles reaches the point where the tortoise started. By the time Achilles gets there, the tortoise has "raced" forward and is now 10 yards ahead of Achilles. (The tortoise moves a tenth as fast as Achilles, so in the time it takes Achilles to travel 100 yards, the tortoise covers a tenth as much ground, or 10 yards.) If you do the math, you find that it took Achilles about 10 seconds to run the 100 yards. (For the sake of argument, let's call it exactly 10 seconds.)

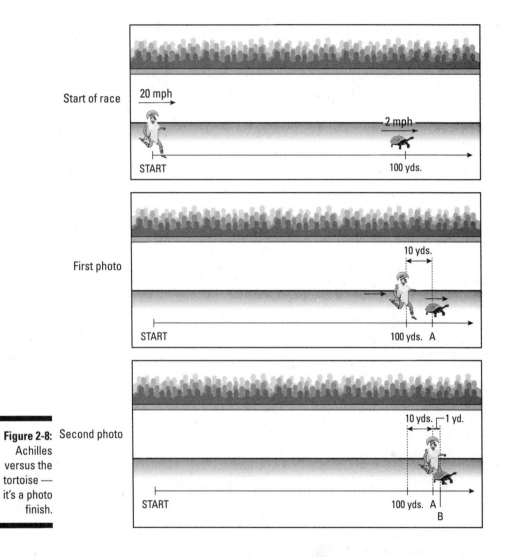

Figure 2-8:
Achilles
versus the
tortoise —
it's a photo
finish.

You have a cool app that allows you to look at your first photo and note precisely where the tortoise is as Achilles crosses the tortoise's starting point. The tortoise's position is shown as point A in the middle image in Figure 2-8. Then you take your second photo when Achilles reaches point A, which takes him about one more second. In that second, the tortoise has moved ahead 1 yard to point B. You take your third photo (not shown) when Achilles reaches point B and the tortoise has moved ahead to point C.

Every time Achilles reaches the point where the tortoise was, you take another photo. There is no end to this series of photographs. Assuming you and your camera can work infinitely fast, you will take an infinite number of photos. And *every single time* Achilles reaches the point where the tortoise was, the tortoise has covered more ground — even if only a millimeter or a millionth of a millimeter. This process never ends, right? Thus, the argument goes, because you can never get to the end of your infinite series of photos, Achilles can never catch or pass the tortoise.

Well, as everyone knows, Achilles does in fact reach and pass the tortoise — thus the paradox. The mathematics of infinite series explains how this infinite series of time intervals sums to a *finite* amount of time — the precise time when Achilles passes the tortoise. Here's the sum for those who are curious:

$$10 \text{ sec.} + 1 \text{ sec.} + 0.1 \text{ sec.} + 0.01 \text{ sec.} + 0.001 \text{ sec.} + \ldots$$

$$= 11.111\ldots \text{ sec., or } 11\frac{1}{9} \text{ seconds.}$$

Achilles passes the tortoise after $11\frac{1}{9}$ seconds at the $111\frac{1}{9}$-yard mark.

Infinite series problems are rich with bizarre, counterintuitive paradoxes. You see more of them in Part V.

Chapter 3

Why Calculus Works

*I*n Chapters 1 and 2, I talk a lot about the process of zooming in on a curve till it looks straight. The mathematics of calculus works because of this basic nature of curves — that they're *locally straight* — in other words, curves are straight at the microscopic level. The earth is round, but to us it looks flat because we're sort of at the microscopic level when compared to the size of the earth. Calculus works because after you zoom in and curves look straight, you can use regular algebra and geometry with them. The zooming-in process is achieved through the mathematics of limits.

The Limit Concept: A Mathematical Microscope

The mathematics of *limits* is the microscope that zooms in on a curve. Here's how a limit works. Say you want the exact slope or steepness of the parabola $y = x^2$ at the point (1, 1). See Figure 3-1.

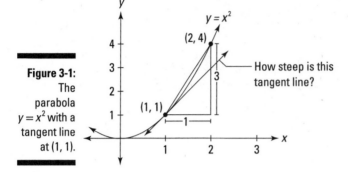

Figure 3-1:
The
parabola
$y = x^2$ with a
tangent line
at (1, 1).

With the slope formula from algebra, you can figure the slope of the line between (1, 1) and (2, 4). From (1, 1) to (2, 4), you go over 1 and up 3, so the slope is $\frac{3}{1}$, or just 3. But you can see in Figure 3-1 that this line is steeper than the tangent line at (1, 1) that shows the parabola's steepness at that specific point. The limit process sort of lets you slide the point that starts at (2, 4) down toward (1, 1) till it's a thousandth of an inch away, then a millionth, then a billionth, and so on down to the microscopic level. If you do the math, the slopes between (1, 1) and your moving point would look something like 2.8, then 2.6, then 2.4, and so on, and then, once you get to a thousandth of an inch away, 2.001, 2.000001, 2.000000001, and so on. And with the almost magical mathematics of limits, you can conclude that the slope at (1, 1) is precisely 2, even though the sliding point never reaches (1, 1). (If it did, you'd only have one point left and you need two separate points to use the slope formula.) The mathematics of limits is all based on this zooming-in process, and it works, again, because the further you zoom in, the straighter the curve gets.

What Happens When You Zoom In

Figure 3-2 shows three diagrams of one curve and three things you might like to know about the curve: 1) the exact slope or steepness at point C, 2) the area under the curve between A and B, and 3) the exact length of the curve from A to B. You can't answer these questions with regular algebra or geometry formulas because the regular formulas for *slope*, *area*, and *length* work for straight lines (and simple curves like circles), but not for weird curves like this one.

The first row of Figure 3-3 shows a magnified detail from the three diagrams of the curve in Figure 3-2. The second row shows further magnification, and the third row yet another magnification. For each little window that gets blown up (like from the first to the second row of Figure 3-3), I've drawn in a new dotted diagonal line to help you see how with each magnification, the blown up pieces of the curves get straighter and straighter. This process is continued indefinitely.

Figure 3-2: One curve — three questions.

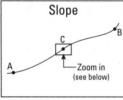

Slope

How steep is the curve at C?

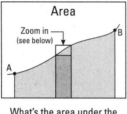

Area

What's the area under the curve between A and B?

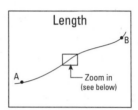

Length

What's the length of the curve from A to B?

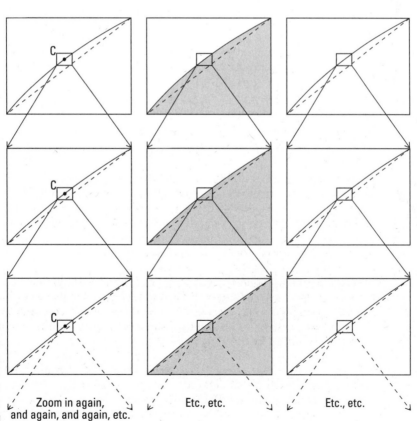

Figure 3-3: Zooming in to the microscopic level.

Zoom in again, and again, and again, etc.　Etc., etc.　Etc., etc.

Finally, Figure 3-4 shows the result after an "infinite" number of magnifications — sort of. After zooming in forever, an infinitely small piece of the original curve and the straight diagonal line are now one and the same. You can think of the lengths 3 and 4 in Figure 3-4 (no pun intended) as 3 and 4 millionths of an inch, no, make that 3 and 4 billionths of an inch, no, trillionths, no, gazillionths, ...

Figure 3-4:
Your final
destina-
tion —
the sub,
sub, sub . . .
subatomic
level.

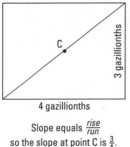

3 gazillionths

4 gazillionths

Slope equals $\frac{rise}{run}$
so the slope at point C is $\frac{3}{4}$.

3 gazillionths

4 gazillionths

The area of a triangle equals
$\frac{1}{2}$ *base* × *height,* so this area is
6 square gazillionths.

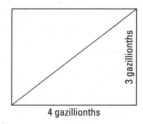

3 gazillionths

4 gazillionths

The Pythagorean theorem
($a^2 + b^2 = c^2$) gives you the length
of the hypotenuse – it's
5 gazillionths.

Now that you've zoomed in "forever," the curve is perfectly straight and you can use regular algebra and geometry formulas to answer the three questions about the curve in Figure 3-2.

For the diagram on the left in Figure 3-4, you can now use the regular *slope* formula from algebra to find the slope at point C. It's exactly $\frac{3}{4}$ — that's the answer to the first question in Figure 3-2. This is how differentiation works.

For the diagram in the middle of Figure 3-4, the regular triangle formula from geometry gives you an area of 6. Then you can get the shaded area inside the strip shown in Figure 3-2 by adding this 6 to the area of the thin rectangle under the triangle (the dark-shaded rectangle in Figure 3-2). Then you repeat this process for all the other narrow strips (not shown), and finally just add up all the little areas. This is how integration works.

And for the diagram on the right of Figure 3-4, the Pythagorean theorem from geometry gives you a length of 5. Then to find the total length of the curve from A to B in Figure 3-2, you do the same thing for the other minute sections of the curve and then add up all the little lengths. This is how you calculate arc length (another integration problem).

Well, there you have it. Calculus uses the limit process to zoom in on a curve till it's straight. After it's straight, the rules of regular-old algebra and geometry apply. Calculus thus gives ordinary algebra and geometry the power to handle complicated problems involving *changing* quantities (which on a graph show up as *curves*). This explains why calculus has so many practical uses, because if there's something you can count on — in addition to death and taxes — it's that things are always changing.

Two Caveats, or Precision, Preschmidgen

Not everything in this chapter (or this book for that matter) will satisfy the high standards of the Grand Poobah of Precision in Mathematical Writing.

I may lose my license to practice mathematics

With regard to the middle diagrams in Figures 3-2 through 3-4, I'm playing a bit fast and loose with the mathematics. The process of integration — finding the area under a curve — doesn't exactly work the way I explained. My explanation isn't really wrong, it's just a bit sideways. But — I don't care what anybody says — that's my story and I'm stickin' to it. Actually, it's not a bad way to think about how integration works, and, anyhow, this is only an introductory chapter.

What the heck does "infinity" really mean?

The second caveat is that whenever I talk about infinity — like in the last section where I discussed zooming in an infinite number of times — I do something like put the word "infinity" in quotes or say something like "you *sort of* zoom in forever." I do this to cover my butt. Whenever you talk about infinity, you're always on shaky ground. What would it mean to zoom in forever or an infinite number of times? You can't do it; you'd never get there. We can imagine — sort of — what it's like to zoom in forever, but there's something a bit fishy about the idea — and thus the qualifications.

Part II
Warming Up with Calculus Prerequisites

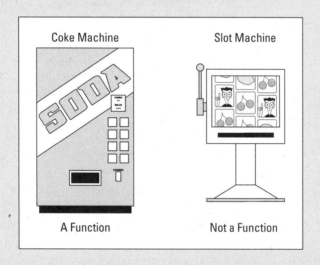

In this part . . .

↙ Algebra review: Richard Feynman, the great 20th century physicist, said (tongue-in-cheek) that calculus was the language that God spoke. Well, I don't know about that, but I do know that algebra is the language of calculus. If you want to learn calculus, you've got to know your algebra.

↙ Logarithm review: What's log 10? And ln 1? Hint for the first one: It's the loneliest number. Hint for the second: There's nothing to it.

↙ Function review: Even and odd functions, exponential functions, inverse functions, function transformations, and so on.

↙ Some trig: The all-important unit circle. And the related geometry of the 30°-60°-90° and 45°-45°-90° triangles.

↙ More trig: SohCahToa and the graphs of sine, cosine, and tangent.

Chapter 4

Pre-Algebra and Algebra Review

. . . .

In This Chapter

▶ Winning the fraction battle: Divide and conquer

▶ Boosting your powers and getting to the root of roots

▶ Laying down the laws of logarithms and having fun with factoring

▶ Hanging around the quad solving quadratics

. . . .

Algebra is the language of calculus. You can't do calculus without knowing algebra any more than you can write Chinese poetry without knowing Chinese. So, if your pre-algebra and algebra are a bit rusty — you know, all those rules for algebraic expressions, equations, fractions, powers, roots, logs, factoring, quadratics, and so on — make sure you review the following basics.

Fine-Tuning Your Fractions

Open a calculus book to any random page and you'll very likely see a fraction — you can't escape them. Dealing with them requires that you know a few rules.

Some quick rules

First is a rule that's simple but very important because it comes up time and time again in the study of calculus:

You can't divide by zero! The denominator of a fraction can *never* equal zero.

$\frac{0}{5}$ equals zero, but $\frac{5}{0}$ is undefined.

It's easy to see why $\frac{5}{0}$ is undefined when you consider how division works:

$$\frac{8}{2} = 4$$

This tells you, of course, that 2 goes into 8 four times; in other words, $2+2+2+2=8$. Well, how many zeros would you need to add up to make 5? You can't do it, and so you can't divide 5 (or any other number) by zero.

Here's another quick rule.

Definition of *reciprocal*: The reciprocal of a number or expression is its multiplicative inverse — which is a fancy way of saying that the product of something and its reciprocal is 1. To get the reciprocal of a fraction, flip it upside down. Thus, the reciprocal of $\frac{3}{4}$ is $\frac{4}{3}$, the reciprocal of 6, which equals $\frac{6}{1}$, is $\frac{1}{6}$, and the reciprocal of $x-2$ is $\frac{1}{x-2}$.

Multiplying fractions

Adding is usually easier than multiplying, but with fractions, the reverse is true — so I want to deal with multiplication first.

Multiplying fractions is a snap — just multiply straight across the top and straight across the bottom:

$$\frac{2}{5} \cdot \frac{3}{4} = \frac{6}{20} = \frac{3}{10} \text{ and } \frac{a}{b} \cdot \frac{c}{d} = \frac{ac}{bd}$$

Dividing fractions

Dividing fractions has one additional step: You flip the second fraction and then multiply — like this:

$$\frac{3}{10} \div \frac{4}{5}$$

$$= \frac{3}{10} \cdot \frac{5}{4} = \frac{15}{40} \quad \text{(Now cancel a 5 from the numerator and denominator.)}$$

$$= \frac{3}{8}$$

Note that you could have canceled before multiplying. Because 5 goes into 5 one time, and 5 goes into 10 two times, you can cancel a 5:

$$\frac{3}{{}_{2}\cancel{10}} \cdot \frac{\cancel{5}^{1}}{4} = \frac{3}{8}$$

Also note that the original problem could have been written as $\dfrac{\frac{3}{10}}{\frac{4}{5}}$.

Adding fractions

You know that

$$\frac{2}{7} + \frac{3}{7} = \frac{2+3}{7} = \frac{5}{7}$$

You can add these up like this because you already have a common denominator. It works the same with variables:

$$\frac{a}{c} + \frac{b}{c} = \frac{a+b}{c}$$

Notice that wherever you have a 2 in the top equation, an a is in the bottom equation; wherever a 3 is in the top equation, a b is in the bottom equation; and ditto for 7 and c. This illustrates a powerful principle:

Variables always behave exactly like numbers.

If you're wondering what to do with variables in a problem, ask yourself how you would do the problem if there were numbers in it instead of variables. Then do the problem with the variables the same way, like this:

$$\frac{a}{b} + \frac{c}{d}$$

You can't add these fractions like you did in the previous example because this problem has no common denominator. Now, assuming you're stumped, do the problem with numbers instead of variables. Remember how to add $\frac{2}{5} + \frac{3}{8}$? I'm not going to simplify each line of the solution. You'll see why in a minute.

1. **Find the *least common denominator* (actually, any common denominator will work when adding fractions), and convert the fractions.**

The least common denominator is 5 times 8, or 40, so convert each fraction into 40ths:

$$\frac{2}{5} + \frac{3}{8}$$

$$= \frac{2}{5} \cdot \frac{8}{8} + \frac{3}{8} \cdot \frac{5}{5}$$

$$= \frac{2 \cdot 8}{5 \cdot 8} + \frac{3 \cdot 5}{5 \cdot 8}$$ (8·5 equals 5·8 so you can reverse the order. These fractions are 40ths, but I want to leave the 5·8 in the denominators for now.)

2. Add the numerators and keep the common denominator unchanged:

$$= \frac{2 \cdot 8 + 3 \cdot 5}{5 \cdot 8} \quad \left(\text{You can see this equals } \frac{16 + 15}{40}, \text{ or } \frac{31}{40}. \right)$$

Now you're ready to do the original problem, $\frac{a}{b} + \frac{c}{d}$. In this problem, you have an a instead of a 2, a b instead of a 5, a c instead of a 3, and a d instead of an 8. Just carry out the exact same steps as you do when adding $\frac{2}{5} + \frac{3}{8}$. You can think of each of the numbers in the above solution as stamped on one side of a coin with the corresponding variable stamped on the other side. For instance, there's a coin with a 2 on one side and an a on the opposite side; another coin has an 8 on one side and a d on the other side, and so on. Now, take each step of the previous solution, flip each coin over, and voilà, you've got the solution to the original problem. Here's the final answer:

$$\frac{ad + cb}{bd}$$

Subtracting fractions

Subtracting fractions works like adding fractions except instead of adding, you subtract. Insights like this are the reason they pay me the big bucks.

Canceling in fractions

Finishing calculus problems — after you've done all the calculus steps — sometimes requires some pretty messy algebra, including canceling. Make sure you know how to cancel and when you can and can't do it.

In the fraction, $\dfrac{x^5 y^2}{x^3 z}$, three xs can be canceled from the numerator and denominator, resulting in the simplified fraction, $\dfrac{x^2 y^2}{z}$. If you write out the xs instead of using exponents, you can more clearly see how this works:

$$\frac{x^5 y^2}{x^3 z} = \frac{x \cdot x \cdot x \cdot x \cdot x \cdot y \cdot y}{x \cdot x \cdot x \cdot z}$$

Now cancel three xs from the numerator and denominator:

$$\frac{\cancel{x} \cdot \cancel{x} \cdot \cancel{x} \cdot x \cdot x \cdot y \cdot y}{\cancel{x} \cdot \cancel{x} \cdot \cancel{x} \cdot z}$$

That leaves you with $\dfrac{x \cdot x \cdot y \cdot y}{z}$, or $\dfrac{x^2 y^2}{z}$.

Express yourself

An *algebraic expression* or just *expression* is something like xyz or $a^2 p^3 \sqrt{q} - 6$, basically anything without an equal sign (if it has an equal sign, it's an *equation*). Canceling works the same way with expressions as it does for single variables. By the way, that's a tip not just for canceling, but for all algebra topics.

Expressions always behave exactly like variables.

So, if each x in the preceding problem is replaced with $(xyz - q)$, you've got

$$\frac{(xyz - q)^5 y^2}{(xyz - q)^3 z}$$

And three of the expression $(xyz - q)$ cancel from the numerator and denominator, just as the three xs canceled. The simplified result is

$$\frac{(xyz - q)^2 y^2}{z}$$

The multiplication rule for canceling

Now you know *how* to cancel. You also need to know *when* you can cancel.

The multiplication rule: You can cancel in a fraction only when it has an *unbroken chain of multiplication* through the entire numerator and the entire denominator.

Canceling is allowed in a fraction like this:

$$\frac{a^2 b^3 (xy - pq)^4 (c+d)}{ab^4 z (xy - pq)^3}$$

Think of multiplication as something that conducts electricity. Electrical current can flow from one end of the numerator to the other, from the a^2 to the $(c+d)$, because all the variables and expressions are connected with multiplication. (Note that an addition or subtraction sign inside parentheses — the "+" in $(c+d)$ for instance — doesn't break the current.) Because the denominator also has an unbroken chain of multiplication, canceling is allowed. You can cancel one a, three bs, and three of the expression $(xy - pq)$. Here's the result:

$$\frac{a(xy - pq)(c+d)}{bz}$$

When you *can't* cancel: But adding an innocuous-looking 1 to the numerator (or denominator) of the original fraction changes everything:

$$\frac{a^2 b^3 (xy - pq)^4 (c+d) + 1}{ab^4 z (xy - pq)^3}$$

The addition sign in front of the 1 breaks the electrical current, and no canceling is allowed anywhere in the fraction.

Absolute Value — Absolutely Easy

Absolute value just turns a negative number into a positive and does nothing to a positive number or zero. For example,

$$|-6| = 6, \quad |3| = 3, \text{ and } |0| = 0$$

It's a bit trickier when dealing with variables. If x is zero or positive, then the absolute value bars do nothing, and thus,

$$|x| = x$$

But if x is negative, the absolute value of x is positive, and you write

$$|x| = -x$$

For example, if $x = -5$, $|-5| = -(-5) = 5$.

$-x$ can be a positive number. When x is a negative number, $-x$ (read as "negative x," or "the opposite of x") is a *positive*.

Empowering Your Powers

You are power*less* in calculus if you don't know the power rules:

- $x^0 = 1$

    This is the rule regardless of what x equals — a fraction, a negative, anything — except for zero (zero raised to the zero power is undefined). Let's call it the kitchen sink rule (where the kitchen sink represents zero): (everything but the kitchen sink)$^0 = 1$

- $x^{-3} = \dfrac{1}{x^3}$ and $x^{-a} = \dfrac{1}{x^a}$

    For example, $4^{-2} = \dfrac{1}{4^2} = \dfrac{1}{16}$. This is huge! Don't forget it! Note that the power is negative, but the answer of $\dfrac{1}{16}$ is *not* negative.

- $x^{2/3} = \left(\sqrt[3]{x}\right)^2 = \sqrt[3]{x^2}$ and $x^{a/b} = \left(\sqrt[b]{x}\right)^a = \sqrt[b]{x^a}$

    You can use this handy rule backwards to convert a root problem into an easier power problem.

- $x^2 \cdot x^3 = x^5$ and $x^a \cdot x^b = x^{a+b}$

    You *add* the powers here. (By the way, you can't do anything to x^2 *plus* x^3. You can't add x^2 to x^3 because they're not *like terms*. You can only add or subtract terms when the variable part of each term is the same, for instance, $3xy^2z + 4xy^2z = 7xy^2z$. This works for exactly the same reason — I'm not kidding — that 3 chairs plus 4 chairs is 7 chairs; and you can't add *unlike* terms, just like you can't add 5 chairs plus 2 cars.)

- $\dfrac{x^5}{x^3} = x^2$ and $\dfrac{x^2}{x^6} = x^{-4}$ and $\dfrac{x^a}{x^b} = x^{a-b}$

    Here you *subtract* the powers.

- $(x^2)^3 = x^6$ and $(x^a)^b = x^{ab}$

    You *multiply* the powers here.

- $(xyz)^3 = x^3 y^3 z^3$ and $(xyz)^a = x^a y^a z^a$

  Here you *distribute* the power to each variable.

- $\left(\dfrac{x}{y}\right)^4 = \dfrac{x^4}{y^4}$ and $\left(\dfrac{x}{y}\right)^a = \dfrac{x^a}{y^a}$

  Here you also distribute the power to each variable.

- $(x + y)^2 = x^2 + y^2$ **NOT!**

  Do *not* distribute the power in this case. Instead, multiply it out the long way: $(x+y)^2 = (x+y)(x+y) = x^2 + xy + xy + y^2 = x^2 + 2xy + y^2$. Watch what happens if you erroneously use the preceding "law" with numbers: $(3+5)^2$ equals 8^2, or 64, *not* $3^2 + 5^2$, which equals $9 + 25$, or 34.

Rooting for Roots

Roots, especially square roots, come up all the time in calculus. So knowing how they work and understanding the fundamental connection between roots and powers is essential. And, of course, that's what I'm about to tell you.

Roots rule — make that, root rules

Any root can be converted into a power, for example, $\sqrt[3]{x} = x^{1/3}$, $\sqrt{x} = x^{1/2}$, and $\sqrt[4]{x^3} = x^{3/4}$. So, if you get a problem with roots in it, you can just convert each root into a power and use the power rules instead to solve the problem (this is a very useful technique). Because you have this option, the following root rules are less important than the power rules, but you really should know them anyway:

- $\sqrt{0} = 0$ and $\sqrt{1} = 1$

  But you knew that, right?

  No negatives under even roots. You can't have a negative number under a square root or under any other *even* number root — at least not in basic calculus.

- $\sqrt{a} \cdot \sqrt{b} = \sqrt{a \cdot b}$, $\sqrt[3]{a} \cdot \sqrt[3]{b} = \sqrt[3]{ab}$, and $\sqrt[n]{a} \cdot \sqrt[n]{b} = \sqrt[n]{ab}$

- $\dfrac{\sqrt{a}}{\sqrt{b}} = \sqrt{\dfrac{a}{b}}$, $\dfrac{\sqrt[3]{a}}{\sqrt[3]{b}} = \sqrt[3]{\dfrac{a}{b}}$, and $\dfrac{\sqrt[n]{a}}{\sqrt[n]{b}} = \sqrt[n]{\dfrac{a}{b}}$

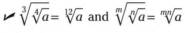

- $\sqrt[3]{\sqrt[4]{a}} = \sqrt[12]{a}$ and $\sqrt[m]{\sqrt[n]{a}} = \sqrt[mn]{a}$

You *multiply* the root indexes.

- $\sqrt{a^2} = |a|$, $\sqrt[4]{a^4} = |a|$, $\sqrt[6]{a^6} = |a|$, **and so on.**

If you have an *even* number root, you need the absolute value bars on the answer, because whether a is positive or negative, the answer is positive. If it's an odd number root, you don't need the absolute value bars. Thus,

- $\sqrt[3]{a^3} = a$, $\sqrt[5]{a^5} = a$, and so on.

- $\sqrt{a^2 + b^2} = a + b$ **NOT!**

Make this mistake and go directly to jail. Try solving it with numbers: $\sqrt{2^2 + 3^2} = \sqrt{13}$, which does *not* equal $2 + 3$.

Simplifying roots

Here are two last things on roots. First, you need to know the two methods for simplifying roots like $\sqrt{300}$ or $\sqrt{504}$.

The quick method works for $\sqrt{300}$ because it's easy to see a large perfect square, 100, that goes into 300. Because 300 equals 100 times 3, the 100 comes out as its square root, 10, leaving the 3 inside the square root. The answer is thus $10\sqrt{3}$.

For $\sqrt{504}$, it's not as easy to find a large perfect square that goes into 504, so you've got to use the longer method:

1. **Break 504 down into a product of all of its prime factors.**

   $$\sqrt{504} = \sqrt{2 \cdot 2 \cdot 2 \cdot 3 \cdot 3 \cdot 7}$$

2. **Circle each pair of numbers.**

   $$\sqrt{\boxed{2 \cdot 2} \cdot 2 \cdot \boxed{3 \cdot 3} \cdot 7}$$

3. **For each circled pair, take one number out.**

   $$2 \cdot 3 \sqrt{2 \cdot 7}$$

4. **Simplify.**

   $$6\sqrt{14}$$

The last thing about roots is that, by convention, you don't leave a root in the denominator of a fraction — it's a silly, anachronistic convention, but it's still being taught, so here it is. If your answer is, say, $\dfrac{2}{\sqrt{3}}$, you multiply it by $\dfrac{\sqrt{3}}{\sqrt{3}}$:

$$\frac{2}{\sqrt{3}} \cdot \frac{\sqrt{3}}{\sqrt{3}} = \frac{2\sqrt{3}}{3}$$

Logarithms — This Is Not an Event at a Lumberjack Competition

A *logarithm* is just a different way of expressing an exponential relationship between numbers. For instance,

$2^3 = 8$, so,

$\log_2 8 = 3$ (read as "log base 2 of 8 equals 3").

These two equations say precisely the same thing. You could think of $2^3 = 8$ as the way we write it in English and $\log_2 8 = 3$ as the way they write it in Latin. And because it's easier to think and do math in English, make sure — when you see something like $\log_3 81 = x$ — that you can instantly "translate" it into $3^x = 81$. The base of a logarithm can be any number greater than zero other than 1, and by convention, if the base is 10, you don't write it. For example, $\log 1000 = 3$ means $\log_{10} 1000 = 3$. Also, log base e ($e \approx 2.72$) is written *ln* instead of $\log_e$.

You should know the following logarithm properties:

- $\log_c 1 = 0$
- $\log_c c = 1$
- $\log_c(ab) = \log_c a + \log_c b$
- $\log_c\left(\dfrac{a}{b}\right) = \log_c a - \log_c b$
- $\log_c a^b = b \log_c a$
- $\log_a b = \dfrac{\log_c b}{\log_c a}$

With this property, you can compute something like $\log_3 20$ on a calculator that only has log buttons for base 10 (the "log" button) and base e (the "ln" button) by entering $\dfrac{\log 20}{\log 3}$, using base 10 for c. On many newer-model calculators, you can compute $\log_3 20$ directly.

✔ $\log_a a^b = b$

✔ $a^{\log_a b} = b$

Factoring Schmactoring — When Am I Ever Going to Need It?

When are you ever going to need it? For calculus, that's when.

Factoring means "unmultiplying," like rewriting 12 as $2 \cdot 2 \cdot 3$. You won't run across problems like that in calculus, however. For calculus, you need to be able to factor algebraic expressions, like factoring $5xy + 10yz$ as $5y(x + 2z)$. Algebraic factoring always involves rewriting a *sum* of terms as a *product*. What follows is a quick refresher course.

Pulling out the GCF

The first step in factoring any type of expression is to pull out — in other words, factor out — the greatest thing that all of the terms have in common — that's the *greatest common factor* or GCF. For example, each of the three terms of $8x^3y^4 + 12x^2y^5 + 20x^4y^3z$ contains the factor $4x^2y^3$, so it can be pulled out like this: $4x^2y^3(2xy + 3y^2 + 5x^2z)$. Make sure you always look for a GCF to pull out before trying other factoring techniques.

Looking for a pattern

After pulling out the GCF if there is one, the next thing to do is to look for one of the following three patterns. The first pattern is *huge*; the next two are much less important.

Difference of squares

Knowing how to factor the *difference of squares* is critical:

$$a^2 - b^2 = (a - b)(a + b)$$

If you can rewrite something like $9x^4 - 25$ so that it looks like $(\text{this})^2 - (\text{that})^2$, then you can use this factoring pattern. Here's how:

$$9x^4 - 25 = (3x^2)^2 - (5)^2$$

Now, because $(\text{this})^2 - (\text{that})^2 = (\text{this} - \text{that})(\text{this} + \text{that})$, you can factor the problem:

$$(3x^2)^2 - (5)^2 = (3x^2 - 5)(3x^2 + 5)$$

A *difference* of squares, $a^2 - b^2$, can be factored, but a *sum* of squares, $a^2 + b^2$, *cannot* be factored. In other words, $a^2 + b^2$, like the numbers 7 and 13, is *prime* — you can't break it up.

Sum and difference of cubes

You might also want to memorize the factor rules for the *sum* and *difference* of cubes:

$$a^3 + b^3 = (a + b)(a^2 - ab + b^2)$$
$$a^3 - b^3 = (a - b)(a^2 + ab + b^2)$$

Trying some trinomial factoring

Remember regular old trinomial factoring from your algebra days?

Several definitions: A *trinomial* is a polynomial with three terms. A *polynomial* is an expression like $4x^5 - 6x^3 + x^2 - 5x + 2$ where, except for the *constant* (the 2 in this example), all the terms have a variable raised to a positive *integral* power. In other words, no fraction powers or negative powers allowed (So, $\frac{1}{x}$ is not a polynomial because it equals x^{-1}). And no radicals, no logs, no sines or cosines, or anything else — just terms with a *coefficient*, like the 4 in $4x^5$, multiplied by a variable raised to a power. The *degree* of a polynomial is the polynomial's highest power of x. The polynomial at the beginning of this paragraph, for instance, has a degree of 5.

It wouldn't be a bad idea to get back up to speed with problems like

$$6x^2 + 13x - 5 = (2x + 5)(3x - 1)$$

where you have to factor the trinomial on the left into the product of the two binomials on the right. A few standard techniques for factoring a trinomial like this are floating around the mathematical ether — you probably learned one or more of them in your algebra class. If you remember one of the techniques, great. You won't have to do a lot of trinomial factoring in calculus, but it does come in handy now and then, so, if your skills are a bit rusty, check out *Algebra II For Dummies* by Mary Jane Sterling (Wiley, 2007).

Solving Quadratic Equations

A quadratic equation is any *second degree* polynomial equation — that's when the highest power of x, or whatever other variable is used, is 2.

You can solve quadratic equations by one of three basic methods.

Method 1: Factoring

Solve $2x^2 - 5x = 12$.

1. **Bring all terms to one side of the equation, leaving a zero on the other side.**

   $$2x^2 - 5x - 12 = 0$$

2. **Factor.**

   $$(2x + 3)(x - 4) = 0$$

   You can check that these factors are correct by multiplying them. Does FOIL (First, Outer, Inner, Last) ring a bell?

3. **Set each factor equal to zero and solve (using the *zero product property*).**

   $$2x + 3 = 0 \qquad \text{or} \qquad x - 4 = 0$$
   $$2x = -3 \qquad\qquad\qquad x = 4$$
   $$x = -\frac{3}{2}$$

So, this equation has two solutions: $x = -\frac{3}{2}$ and $x = 4$.

The *discriminant* **tells you whether a quadratic is factorable.** Method 1 will work only if the quadratic is factorable. The quick test for that is a snap. A quadratic is factorable if the discriminant, $b^2 - 4ac$, is a perfect square number like 0, 1, 4, 9, 16, 25, etc. (The discriminant is the stuff under the square root symbol in the quadratic formula — see Method 2 below.) In the quadratic equation from Step 1 above, $2x^2 - 5x - 12 = 0$, for example, $a = 2$, $b = -5$, and $c = -12$. $b^2 - 4ac$ equals, therefore, $(-5)^2 - 4(2)(-12)$ which equals 121. Since 121 is a perfect square (11^2), the quadratic is factorable. Because trinomial factoring is often so quick and easy, you may choose to just dive into the problem and try to factor it without bothering to check the value of the discriminant. But if you get stuck, it's not a bad idea to check the discriminant so you don't waste more time trying to factor an unfactorable quadratic trinomial. (But whether or not the quadratic is factorable, you can always solve it with the quadratic formula discussed in the next section.)

Method 2: The quadratic formula

The solution or solutions of a quadratic equation, $ax^2 + bx + c = 0$, are given by the quadratic formula:

$$x = \frac{-b \pm \sqrt{b^2 - 4ac}}{2a}$$

Now solve the same equation from Method 1 with the quadratic formula:

1. **Bring all terms to one side of the equation, leaving a zero on the other side.**

   $$2x^2 - 5x - 12 = 0$$

2. **Plug the coefficients into the formula.**

   In this example, a equals 2, b is –5, and c is –12, so

   $$x = \frac{-(-5) \pm \sqrt{(-5)^2 - 4(2)(-12)}}{2 \cdot 2}$$

   $$= \frac{5 \pm \sqrt{25 - (-96)}}{4}$$

   $$= \frac{5 \pm \sqrt{121}}{4}$$

   $$= \frac{5 \pm 11}{4}$$

   $$= \frac{16}{4} \text{ or } -\frac{6}{4}$$

   $$x = 4 \text{ or } -\frac{3}{2}$$

   This agrees with the solutions obtained previously — the solutions better be the same because we're solving the same equation.

Method 3: Completing the square

The third method of solving quadratic equations is called *completing the square* because it involves creating a perfect square trinomial that you can solve by taking its square root.

Solve $3x^2 = 24x + 27$.

1. **Put the x^2 and the x terms on one side and the constant on the other.**

   $3x^2 - 24x = 27$

2. **Divide both sides by the coefficient of x^2 (unless, of course, it's 1).**

   $x^2 - 8x = 9$

3. **Take half of the coefficient of x, square it, then add that to both sides.**

   Half of -8 is -4 and $(-4)^2$ is 16, so add 16 to both sides:

   $x^2 - 8x + 16 = 9 + 16$

4. **Factor the left side into a binomial squared. Notice that the factor always contains the same number you found in Step 3 (–4 in this example).**

   $(x - 4)^2 = 25$

5. **Take the square root of both sides, remembering to put a $\pm$ sign on the right side.**

   $$\sqrt{(x-4)^2} = \sqrt{25}$$
   $$x - 4 = \pm 5$$

6. **Solve.**

   $x = 4 \pm 5$
   $x = 9 \text{ or } -1$

Chapter 5

Funky Functions and Their Groovy Graphs

● ●

In This Chapter

▶ Figuring out functions and relations

▶ Learning about lines

▶ Getting particular about parabolas

▶ Grappling with graphs

▶ Transforming functions and investigating inverse functions

● ●

*V*irtually everything you do in calculus concerns functions and their graphs in one way or another. Differential calculus involves finding the slope or steepness of various functions, and integral calculus involves computing the area underneath functions. And not only is the concept of a function critical for calculus, it's one of the most fundamental ideas in all of mathematics.

What Is a Function?

Basically, a function is a relationship between two things in which the numerical value of one thing in some way depends on the value of the other. Examples are all around us: The average daily temperature for your city depends on, and is a function of, the time of year; the distance an object has fallen is a function of how much time has elapsed since you dropped it; the area of a circle is a function of its radius; and the pressure of an enclosed gas is a function of its temperature.

The defining characteristic of a function

A function has only one output for each input.

Consider Figure 5-1.

Coke Machine   Slot Machine

Figure 5-1:
The Coke
machine is
a function.
The slot
machine
is not.

A Function   Not a Function

The Coke machine is a function because after plugging in the inputs (your choice and your money), you know exactly what the output is. With the slot machine, on the other hand, the output is a mystery, so it's not a function. Look at Figure 5-2.

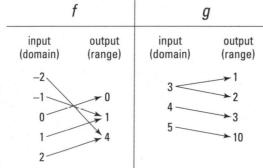

Figure 5-2:
f is a
function;
g is not.

The squaring function, *f*, is a function because it has exactly one output assigned to each input. It doesn't matter that both 2 and −2 produce the same output of 4 because given an input, say −2, there's no mystery about the output. When you input 3 into *g*, however, you don't know whether the output is 1 or 2. (For now, don't worry about how the *g* rule turns its inputs into its outputs.) Because no output mysteries are allowed in functions, *g* is not a function.

Good functions, unlike good literature, have predictable endings.

Definitions of *domain* and *range*: The set of all inputs of a function is called the domain of the function; the set of all outputs is the range of the function.

Some people like to think of a function as a machine. Consider again the squaring function, *f*, from Figure 5-2. Figure 5-3 shows two of the inputs and their respective outputs.

Figure 5-3:
A function
machine:
Meat goes
in, sausage
comes out.

f

1
−2

x^2

1
4

You pop a 1 into the function machine, and out pops a 1; you put in a −2 and a 4 comes out. A function machine takes an input, operates on it in some way, then spits out the output.

Independent and dependent variables

Definitions of *dependent variable* and *independent variable*: In a function, the thing that *depends* on the other thing is called the *dependent variable*; the other thing is the *independent variable*. Because you plug numbers into the independent variable, it's also called the *input variable*. After plugging in a number, you then calculate the output or answer for the dependent variable, so the dependent variable is also called the *output variable*. When you graph a function, the independent variable goes on the *x*-axis, and the dependent variable goes on the *y*-axis.

Sometimes the dependence between the two things is one of cause and effect — for example, raising the temperature of a gas *causes* an increase in the pressure. In this case, temperature is the *independent variable* and pressure the *dependent variable* because the pressure depends on the temperature.

Often, however, the dependence is not one of cause and effect, but just some sort of association between the two things. Usually, though, the independent variable is the thing we already know or can easily ascertain, and the dependent variable is the thing we want to figure out. For instance, you wouldn't say that time causes an object to fall (gravity is the cause), but if you know how much time has passed since you dropped an object, you can figure out how far it has fallen. So, time is the independent variable, and distance fallen is the dependent variable; and you would say that distance is a function of time.

Whatever the type of correspondence between the two variables, the dependent variable (the y-variable) is the thing we're usually more interested in. Generally, we want to know what happens to the dependent or y-variable as the independent or x-variable goes to the right: Is the y-variable (the height of the graph) rising or falling and, if so, how steeply, or is the graph level, neither going up nor down?

Function notation

A common way of writing the function $y = 5x^3 - 2x^2 + 3$ is to replace the "y" with "$f(x)$" and write $f(x) = 5x^3 - 2x^2 + 3$. It's just a different notation for the same thing. These two equations are, in every respect, mathematically identical. Students are often puzzled by function notation when they see it the first time. They wonder what the "f" is and whether $f(x)$ means f times x. It does not. If function notation bugs you, my advice is to think of $f(x)$ as simply the way y is written in some foreign language. Don't consider the f and the x separately; just think of $f(x)$ as a single symbol for y.

You can also think of $f(x)$ (read as "f of x") as short for "a function of x." You can write $y = f(x) = 3x^2$, which is translated as "y is a function of x and that function is $3x^2$." However, sometimes other letters are used instead of f — such as $g(x)$ or $p(x)$ — often just to differentiate between functions. The function letter doesn't necessarily stand for anything, but sometimes the initial letter of a word is used (in which case you use an uppercase letter). For instance, you know that the area of a square is determined by squaring the length of its side: $Area = side^2$ or $A = s^2$. The area of a square depends on, and is a *function* of, the length of its side. With function notation, you can write $A(s) = s^2$. (Quick quiz: How does $f(x) = x^2$ differ from the area of a square function, $A(s) = s^2$? Answer: for $f(x) = x^2$, x can equal any number, but with $A(s) = s^2$, s must be positive, because the length of a side of a square cannot be negative or zero. The two functions thus have different *domains*.)

Consider, again, the squaring function $y = x^2$ or $f(x) = x^2$. When you input 3 for x, the output is 9. Function notation is convenient because you can concisely express the input and the output by writing $f(3) = 9$ (read as "f of 3 equals 9"). Remember that $f(3) = 9$ means that when x is 3, $f(x)$ is 9; or, equivalently, it tells you that when x is 3, y is 9.

Composite functions

A *composite* function is the combination of two functions. For example, the cost of the electrical energy needed to air condition your place depends on how much electricity you use, and usage depends on the outdoor temperature.

Because cost depends on usage and usage depends on temperature, cost will depend on temperature. In function language, cost is a function of usage, usage is a function of temperature, and thus cost is a function of temperature. This last function, a combination of the first two, is a composite function.

Let $f(x)=x^2$ and $g(x)=5x-8$. Input 3 into $g(x)$: $g(3)=5\cdot3-8$, which equals 7. Now take that output, 7, and plug it into $f(x)$: $f(7)=7^2=49$. The machine metaphor shows what I did here. Look at Figure 5-4. The g machine turns the 3 into a 7, and then the f machine turns the 7 into a 49.

Figure 5-4:
Two
function
machines.

You can express the net result of the two functions in one step with the following *composite* function:

$$f(g(3))=49$$

You always calculate the inside function of a composite function first: $g(3)=7$. Then you take the output, 7, and calculate $f(7)$, which equals 49.

To determine the general composite function, $f(g(x))$, plug $g(x)$, which equals $5x-8$, into $f(x)$. In other words, you want to determine $f(5x-8)$. The f function or f machine takes an input and squares it. Thus,

$$f(5x-8)=(5x-8)^2$$
$$=(5x-8)(5x-8)$$
$$=25x^2-40x-40x+64$$
$$=25x^2-80x+64$$

Thus, $f(g(x))=25x^2-80x+64$.

With composite functions, the order matters. As a general rule, $f(g(x))\neq g(f(x))$.

What Does a Function Look Like?

I'm no math historian, but everyone seems to agree that René Descartes (1596–1650) came up with the x-y coordinate system shown in Figure 5-5.

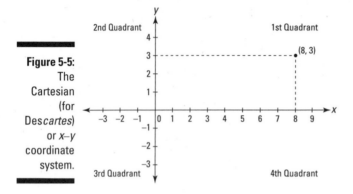

Figure 5-5:
The Cartesian (for Des*cartes*) or x–y coordinate system.

Isaac Newton (1642–1727) and Gottfried Leibniz (1646–1716) are credited with inventing calculus, but it's hard to imagine that they could have done it without Descartes' contribution several decades earlier. Think of the coordinate system (or the screen on your graphing calculator) as your window into the world of calculus. Virtually everything in your calculus textbook and in this book involves (directly or indirectly) the graphs of lines or curves — usually functions — in the x-y coordinate system.

Consider the four graphs in Figure 5-6.

These four curves are functions because they satisfy the *vertical line test*. (***Note:*** I'm using the term *curve* here to refer to any shape, whether it's curved or straight.)

The vertical line test: A curve is a function if a vertical line drawn through the curve — regardless of where it's drawn — touches the curve only once. This guarantees that each input within the function's domain has exactly one output.

No matter where you draw a vertical line on any of the four graphs in Figure 5-6, the line touches the curve at only one point. Try it.

If, however, a vertical line can be drawn so that it touches a curve two or more times, then the curve is not a function. The two curves in Figure 5-7, for example, are not functions.

$y = 3x + 5$

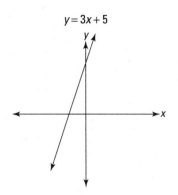

$y = x^2 - 2$

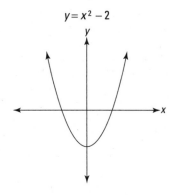

$y = |x|$

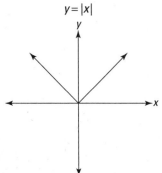

$y = \sin x$

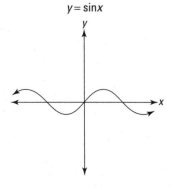

Figure 5-6:
Four
functions.

(Note: These graphs have different scales.)

Figure 5-7:
These two
curves
are not
functions
because
they fail
the vertical
line test.
They are,
however,
relations.

$x^2 + y^2 = 9$

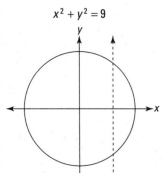

$x = y^3 - 5y^2 + 10$

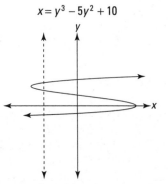

(Note: These graphs have different scales.)

So, the four curves in Figure 5-6 are functions, and the two in Figure 5-7 are not, but all six of the curves are *relations*.

Definition of *relation*: A relation is any collection of points on the *x-y* coordinate system.

You spend a little time studying some non–function relations in calculus — circles, for instance — but the vast majority of calculus problems involve functions.

Common Functions and Their Graphs

You're going to see hundreds of functions in your study of calculus, so it wouldn't be a bad idea to familiarize yourself with the basic ones in this section: the line, the parabola, the absolute value function, the cubing and cube root functions, and the exponential and logarithmic functions.

Lines in the plane in plain English

A line is the simplest function you can graph on the coordinate plane. (Lines are important in calculus because you often study lines that are tangent to curves and because when you zoom in far enough on a curve, it looks and behaves like a line.) Figure 5-8 shows an example of a line: $y = 3x + 5$.

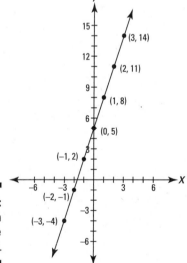

Figure 5-8:
The graph of the line $y = 3x + 5$.

Hitting the slopes

The most important thing about the line in Figure 5-8 — at least for your study of calculus — is its slope or steepness. Notice that whenever x goes 1 to the right, y goes up by 3. A good way to visualize slope is to draw a stairway under the line (see Figure 5-9). The vertical part of the step is called the *rise*, the horizontal part is called the *run*, and the slope is defined as the ratio of the rise to the run:

$$Slope = \frac{rise}{run} = \frac{3}{1} = 3$$

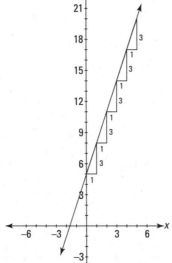

Figure 5-9:
The line
$y = 3x + 5$
has a slope
of 3.

You don't have to make the run equal to 1. The ratio of rise to run, and thus the slope, always comes out the same, regardless of what size you make the steps. If you make the run equal to 1, however, the slope is the same as the rise because a number divided by 1 equals itself. This is a good way to think about slope — the slope is the amount that a line goes up (or down) as it goes 1 to the right.

Definitions of positive, negative, zero, and undefined slopes: Lines that go *up* to the right have a *positive* slope; lines that go *down* to the right have a *negative* slope. Horizontal lines have a slope of *zero*, and vertical lines do not have a slope — you say that the slope of a vertical line is *undefined*.

Here's the formula for slope:

$$Slope = \frac{y_2 - y_1}{x_2 - x_1}$$

Pick any two points on the line in Figure 5-9, say $(1, 8)$ and $(3, 14)$, and plug them into the formula to calculate the slope:

$$Slope = \frac{14 - 8}{3 - 1} = \frac{6}{2} = 3$$

This computation involves, in a sense, a stairway step that goes over 2 and up 6. The answer of 3 agrees with the slope you can see in Figure 5-9.

Any line parallel to this one has the same slope, and any line perpendicular to this one has a slope of $-\frac{1}{3}$, which is the *opposite reciprocal* of 3.

Parallel lines have the same slope. Perpendicular lines have opposite reciprocal slopes.

Graphing lines

If you have the equation of the line, $y = 3x + 5$, but not its graph, you can graph the line the old-fashioned way or with your graphing calculator. The old-fashioned way is to create a table of values by plugging numbers into x and calculating y. If you plug 0 into x, y equals 5; plug 1 into x, and y equals 8; plug 2 into x, and y is 11, and so on. Table 5-1 shows the results.

Table 5-1	Points on the Line $y = 3x + 5$					
x	0	1	2	3	4	------▶
y	5	8	11	14	17	------▶

Plot the points, connect the dots, and put arrows on both ends — there's your line. This is a snap with a graphing calculator. Just enter $y = 3x + 5$ and your calculator graphs the line and produces a table like Table 5-1.

Slope-intercept and point-slope forms

You can see that the line in Figure 5-9 crosses the y-axis at 5 — this point is the *y-intercept* of the line. Because both the slope of 3 and the y-intercept of 5 appear in the equation $y = 3x + 5$, this equation is said to be in *slope-intercept* form. Here's the form written in the general way:

***Slope-intercept* form:**

$$y = mx + b$$

(Where m is the slope and b is the y-intercept.)

(If that doesn't ring a bell — even a distant, faint bell — go directly to the registrar and drop calculus, but do *not* under any circumstances return this book.)

All lines, except for *vertical* lines, can be written in this form. Vertical lines are written like $x = 6$, for example. The number tells you where the vertical line crosses the x–axis.

The equation of a *horizontal* line also looks different, $y = 10$ for example. But it technically fits the form $y = mx + b$ — it's just that because the slope of a horizontal line is zero, and because zero times x is zero, there is no x-term in the equation. (But, if you felt like it, you could write $y = 10$ as $y = 0x + 10$.)

Definition of a *constant function*: A line is the simplest type of function, and a horizontal line (called a constant function) is the simplest type of line. It's nonetheless fairly important in calculus, so make sure you know that a horizontal line has an equation like $y = 10$ and that its slope is zero.

If $m = 1$ and $b = 0$, you get the function $y = x$. This line goes through the *origin* $(0,0)$ and makes a $45°$ angle with both coordinate axes. It's called the *identity* function because its outputs are the same as its inputs.

***Point-slope* form:** In addition to the slope-intercept form for the equation of a line, you should also know the point-slope form:

$$y - y_1 = m(x - x_1)$$

To use this form, you need to know — you guessed it — a *point* on a line and the line's *slope*. You can use any point on the line. Consider the line in Figure 5-9 again. Pick any point, say $(2, 11)$, and then plug the x- and y-coordinates of the point into x_1 and y_1, and plug the slope, 3, into m:

$$y - 11 = 3(x - 2)$$

With a little algebra, you can convert this equation into the one we already know, $y = 3x + 5$. Try it.

Parabolic and absolute value functions — even steven

You should be familiar with the two functions shown in Figure 5-10: the parabola, $f(x)=x^2$, and the absolute value function, $g(x)=|x|$.

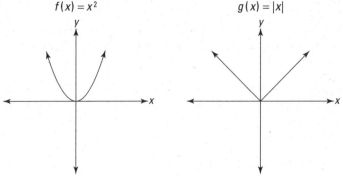

Figure 5-10: The graphs of $f(x)=x^2$ and $g(x)=|x|$.

$$f(x) = x^2 \qquad\qquad g(x) = |x|$$

Notice that both functions are symmetric with respect to the y-axis. In other words, the left and right sides of each graph are mirror images of each other. This makes them *even* functions. A *polynomial* function like $y=9x^4-4x^2+3$, where all powers of x are even, is one type of even function. (Such an even polynomial function can contain — but need not contain — a constant term like the 3 in the preceding function. This makes sense because 3 is the same as $3x^0$ and zero is an even number.) Another even function is $y=\cos(x)$ (see Chapter 6).

A couple oddball functions

Graph $f(x)=x^3$ and $g(x)=\sqrt[3]{x}$ on your graphing calculator. These two functions illustrate *odd* symmetry. Odd functions are symmetric with respect to the origin which means that if you were to rotate them 180° about the origin, they would land on themselves. A polynomial function like $y=4x^5-x^3+2x$, where all powers of x are odd, is one type of odd function. (Unlike an even polynomial function, an odd polynomial function *cannot* contain a constant term.) Another odd function is $y=\sin(x)$ (see Chapter 6).

Many functions are neither even nor odd, for example $y = 3x^2 - 5x$. My high school English teacher said a paragraph should never have just one sentence, so voilà, now it's got two.

Exponential functions

An exponential function is one with a power that contains a variable, such as $f(x) = 2^x$ or $g(x) = 10^x$. Figure 5-11 shows the graphs of both these functions on the same *x-y* coordinate system.

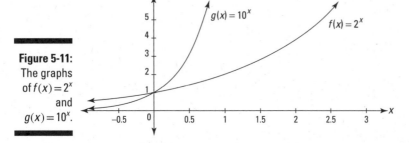

Figure 5-11:
The graphs
of $f(x) = 2^x$
and
$g(x) = 10^x$.

Both functions go through the point $(0, 1)$, as do all exponential functions of the form $f(x) = b^x$. When b is greater than 1, you have *exponential growth*. All such functions get higher and higher without limit as they go to the right toward positive infinity. As they go to the left toward negative infinity, they crawl along the *x*-axis, always getting closer to the axis, but never touching it. You use these and related functions for figuring things like investments, inflation, and growing population.

When b is between 0 and 1, you have an *exponential decay* function. The graphs of such functions are like exponential growth functions in reverse. Exponential decay functions also cross the *y*-axis at $(0, 1)$, but they go up to the *left* forever, and crawl along the *x*-axis to the *right*. These functions model things that shrink over time, such as the radioactive decay of uranium.

Logarithmic functions

A logarithmic function is simply an exponential function with the *x*- and *y*-axes switched. In other words, the up-and-down direction on an exponential graph corresponds to the right-and-left direction on a logarithmic graph, and the right-and-left direction on an exponential graph corresponds to the up-and-down direction on a logarithmic graph. (If you want a refresher on logs, see Chapter 4.) You can see this relationship in Figure 5-12, in which both $f(x)=2^x$ and $g(x)=\log_2 x$ are graphed on the same set of axes.

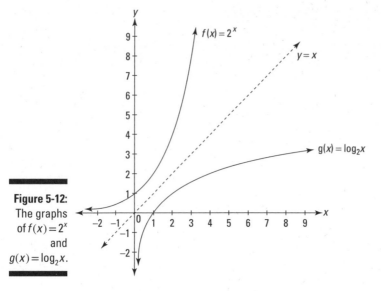

Figure 5-12:
The graphs
of $f(x)=2^x$
and
$g(x)=\log_2 x$.

Both exponential and logarithmic functions are *monotonic*. A monotonic function either goes up over its entire domain (called an *increasing* function) or goes down over its whole domain (a *decreasing* function). (I'm assuming here — as is almost always the case — that the motion along the function is from left to right.)

Notice the symmetry of the two functions in Figure 5-12 about the line $y=x$. This makes them *inverses* of each other, which brings us to the next topic.

Inverse Functions

The function $f(x)=x^2$ (for $x\geq0$) and the function $f^{-1}(x)=\sqrt{x}$ (read as "f inverse of x") are inverse functions because each undoes what the other does. In other words, $f(x)=x^2$ takes an input of, say, 3 and produces an output of 9 (because $3^2=9$); $f^{-1}(x)=\sqrt{x}$ takes the 9 and turns it back into the 3 (because $\sqrt{9}=3$). Notice that $f(3)=9$ and $f^{-1}(9)=3$. You can write all of this in one step as $f^{-1}(f(3))=3$. It works the same way if you start with $f^{-1}(x)$. $f^{-1}(16)=4$ (because $\sqrt{16}=4$), and $f(4)=16$ (because $4^2=16$). If you write this in one step, you get $f(f^{-1}(16))=16$. (Note that while only $f^{-1}(x)$ is read as f *inverse* of x, both functions are inverses of each other.)

The inverse function rule: The fancy way of summing up all of this is to say that $f(x)$ and $f^{-1}(x)$ are inverse functions if and only if $f^{-1}(f(x))=x$ and $f(f^{-1}(x))=x$.

Don't confuse the superscript –1 in $f^{-1}(x)$ with the exponent –1. The exponent –1 gives you the reciprocal of something, for example $x^{-1}=\dfrac{1}{x}$. But $f^{-1}(x)$ is the inverse of $f(x)$. It does *not* equal $\dfrac{1}{f(x)}$, which is the reciprocal of $f(x)$. So why is the exact same symbol used for two different things? Beats me.

When you graph inverse functions, each is the mirror image of the other, reflected over the line $y=x$. Look at Figure 5-13, which graphs the inverse functions $f(x)=x^2$ (for $x\geq0$) and $f^{-1}(x)=\sqrt{x}$.

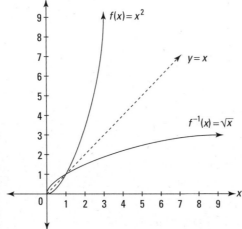

Figure 5-13: The graphs of $f(x)=x^2$, $(x\geq0)$, and $f^{-1}(x)=\sqrt{x}$.

If you rotate the graph in Figure 5-13 counterclockwise so that the line $y=x$ is vertical, you can easily see that $f(x)$ and $f^{-1}(x)$ are mirror images of each other. One consequence of this symmetry is that if a point like (2, 4) is on one of the functions, the point (4, 2) will be on the other. Also, the domain of f is the range of f^{-1}, and the range of f is the domain of f^{-1}.

Shifts, Reflections, Stretches, and Shrinks

Any function can be transformed into a related function by shifting it horizontally or vertically, flipping it over horizontally or vertically, or stretching or shrinking it horizontally or vertically. I do the horizontal transformations first. Consider the exponential function $y = 2^x$. See Figure 5-14.

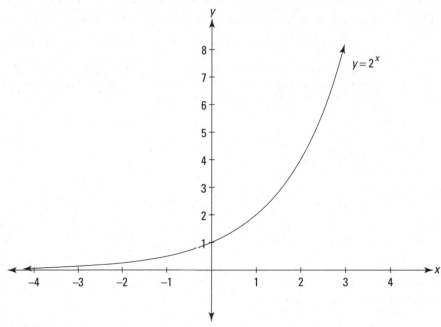

Figure 5-14: The graph of $y = 2^x$.

Horizontal transformations

Horizontal changes are made by adding a number to or subtracting a number from the input variable x or by multiplying x by some number. All horizontal transformations, except reflection, work the *opposite* way you'd expect: Adding to x makes the function go left, subtracting from x makes the function go right, multiplying x by a number greater than 1 shrinks the function, and multiplying x by a number less than 1 expands the function. For example, the graph of $y = 2^{x+3}$ has the same shape and orientation as the graph in

Figure 5-14; it's just shifted three units to the *left*. Instead of passing through $(0, 1)$ and $(1, 2)$, the shifted function goes through $(-3, 1)$ and $(-2, 2)$. And the graph of $y = 2^{x-3}$ is three units to the *right* of $y = 2^x$. The original function and both transformations are shown in Figure 5-15.

If you multiply the x in $y = 2^x$ by 2, the function shrinks horizontally by a factor of 2. So every point on the new function is half of its original distance from the y-axis. The y-coordinate of every point stays the same; the x-coordinate is cut in half. For example, $y = 2^x$ goes through $(1, 2)$, so $y = 2^{2x}$ goes through $(\frac{1}{2}, 2)$; $y = 2^x$ goes through $(-4, \frac{1}{16})$, so $y = 2^{2x}$ goes through $(-2, \frac{1}{16})$. Multiplying x by a number less than 1 has the opposite effect. When $y = 2^x$ is transformed into $y = 2^{(1/4)x}$, every point on $y = 2^x$ is pulled away from the y-axis to a distance 4 times what it was. To visualize the graph of $y = 2^{(1/4)x}$, imagine you've got the graph of $y = 2^x$ on an elastic coordinate system. Grab the coordinate system on the left and right and stretch it by a factor of 4, pulling everything away from the y-axis, but keeping the y-axis in the center. Now you've got the graph of $y = 2^{(1/4)x}$. Check these transformations out on your graphing calculator.

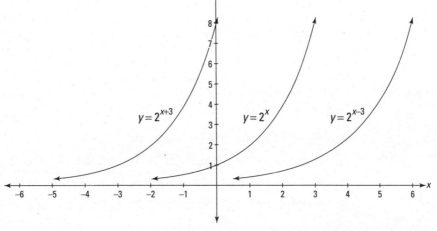

Figure 5-15:
The graphs
of $y = 2^x$,
$y = 2^{x+3}$,
and $y = 2^{x-3}$.

The last horizontal transformation is a reflection over the y-axis. Multiplying the x in $y = 2^x$ by -1 reflects it over or flips it over the y-axis. For instance, the point $(1, 2)$ becomes $(-1, 2)$ and $(-2, \frac{1}{4})$ becomes $(2, \frac{1}{4})$. See Figure 5-16.

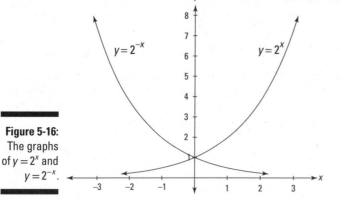

Vertical transformations

To transform a function vertically, you add a number to or subtract a number from the entire function or multiply the whole function by a number. To do something to an entire function, say $y = 10^x$, imagine that the entire right side of the equation is inside parentheses, like $y = (10^x)$. Now, all vertical transformations are made by placing a number somewhere on the right side of the equation *outside* the parentheses. (Often, you don't actually need the parentheses, but sometimes you do.) Unlike horizontal transformations, vertical transformations work the way you expect: Adding makes the function go up, subtracting makes it go down, multiplying by a number greater than 1 stretches the function, and multiplying by a number less than 1 shrinks the function. For example, consider the following transformations of the function $y = 10^x$:

$y = 10^x + 6$ shifts the original function up 6 units.

$y = 10^x - 2$ shifts the original function down 2 units.

$y = 5 \cdot 10^x$ stretches the original function vertically by a factor of 5.

$y = \frac{1}{3} \cdot 10^x$ shrinks the original function vertically by a factor of 3.

Multiplying the function by –1 reflects it over the *x*-axis, or, in other words, flips it upside down. Look at these transformations on your graphing calculator.

As you saw in the previous section, horizontal transformations change only the *x*-coordinates of points, leaving the *y*-coordinates unchanged. Conversely, vertical transformations change only the *y*-coordinates of points, leaving the *x*-coordinates unchanged.

Chapter 6

The Trig Tango

*M*any calculus problems involve trigonometry, and the calculus itself is enough of a challenge without having to relearn trig at the same time. So, if your trig is rusty — I'm shocked — review these trig basics, or else!

Studying Trig at Camp SohCahToa

The study of trig begins with the right triangle. The three main trig functions (sine, cosine, and tangent) and their reciprocals (cosecant, secant, and cotangent) all tell you something about the lengths of the sides of a right triangle that contains a given acute angle — like angle *x* in Figure 6-1. The longest side of this right triangle (or any right triangle), the diagonal side, is called the *hypotenuse*. The side that's 3 units long in this right triangle is referred to as the *opposite* side because it's on the opposite side of the triangle from angle *x*, and the side of length 4 is called the *adjacent* side because it's adjacent to, or touching, angle *x*.

Figure 6-1:
Sitting around the campfire, studying a right triangle.

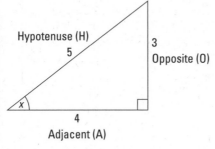

SohCahToa is a meaningless mnemonic device that helps you remember the definitions of the sine, cosine, and tangent functions. *SohCahToa* uses the initial letters of *sine, cosine,* and *tangent,* and the initial letters of *hypotenuse, opposite,* and *adjacent* to help you remember the following definitions. (To remember how to spell *SohCahToa,* note its pronunciation and the fact that it contains three groups of three letters each.) For any angle, θ,

Soh

$$\sin \theta = \frac{O}{H}$$

Cah

$$\cos \theta = \frac{A}{H}$$

Toa

$$\tan \theta = \frac{O}{A}$$

For the triangle in Figure 6-1,

$$\sin x = \frac{O}{H} = \frac{3}{5}$$

$$\cos x = \frac{A}{H} = \frac{4}{5}$$

$$\tan x = \frac{O}{A} = \frac{3}{4}$$

The other three trig functions are reciprocals of these: Cosecant (csc) is the reciprocal of sine, secant (sec) is the reciprocal of cosine, and cotangent (cot) is the reciprocal of tangent.

$$\csc \theta = \frac{1}{\sin \theta} = \frac{1}{\frac{O}{H}} = \frac{H}{O} \qquad \sec \theta = \frac{1}{\cos \theta} = \frac{1}{\frac{A}{H}} = \frac{H}{A} \qquad \cot \theta = \frac{1}{\tan \theta} = \frac{1}{\frac{O}{A}} = \frac{A}{O}$$

So for the triangle in Figure 6-1,

$$\csc x = \frac{H}{O} = \frac{5}{3}$$

$$\sec x = \frac{H}{A} = \frac{5}{4}$$

$$\cot x = \frac{A}{O} = \frac{4}{3}$$

Two Special Right Triangles

Because so many garden variety calculus problems involve 30°, 45°, and 60° angles, it's a good idea to memorize the two right triangles in Figure 6-2.

Figure 6-2: The 45°-45°-90° triangle and the 30°-60°-90° triangle.

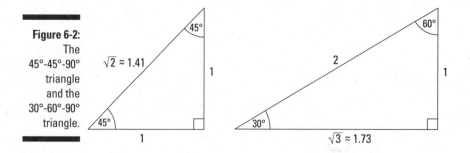

The 45°-45°-90° triangle

Every 45°-45°-90° is the shape of a square cut in half along its diagonal. The 45°-45°-90° triangle in Figure 6-2 is half of a 1-by-1 square. The Pythagorean theorem gives you the length of its hypotenuse, $\sqrt{2}$, or about 1.41.

The Pythagorean theorem: For any right triangle, $a^2 + b^2 = c^2$, where a and b are the lengths of the triangle's *legs* (the sides touching the right angle) and c is the length of its *hypotenuse*.

When you apply the *SohCahToa* trig functions and their reciprocals to the 45° angle in the 45°-45°-90° triangle, you get the following trig values:

$$\sin 45° = \frac{O}{H} = \frac{1}{\sqrt{2}} = \frac{\sqrt{2}}{2} \approx 0.71 \qquad \csc 45° = \frac{H}{O} = \frac{\sqrt{2}}{1} = \sqrt{2} \approx 1.41$$

$$\cos 45° = \frac{A}{H} = \frac{1}{\sqrt{2}} = \frac{\sqrt{2}}{2} \approx 0.71 \qquad \sec 45° = \frac{H}{A} = \frac{\sqrt{2}}{1} = \sqrt{2} \approx 1.41$$

$$\tan 45° = \frac{O}{A} = \frac{1}{1} = 1 \qquad \cot 45° = \frac{A}{O} = \frac{1}{1} = 1$$

The 30°-60°-90° triangle

Every 30°-60°-90° triangle is half of an equilateral triangle cut straight down the middle along its altitude.

The 30°-60°-90° in Figure 6-2 is half of a 2-by-2-by-2 equilateral triangle. It has legs of lengths 1 and $\sqrt{3}$ (about 1.73), and a 2-unit long hypotenuse.

Don't make the common error of switching the 2 with the $\sqrt{3}$ in a 30°-60°-90° triangle. Remember that 2 is more than $\sqrt{3}$ ($\sqrt{4}$ equals 2, so $\sqrt{3}$ be must be less than 2) and that the hypotenuse is always the longest side of a right triangle.

When you sketch a 30°-60°-90° triangle, exaggerate the fact that it's wider than it is tall (or taller than wide if you tip it up). This makes it obvious that the shortest side (length of 1) is opposite the smallest angle (30°).

Here are the trig values for the 30° angle in the 30°-60°-90° triangle:

$$\sin 30° = \frac{O}{H} = \frac{1}{2} \qquad \csc 30° = \frac{H}{O} = \frac{2}{1} = 2$$

$$\cos 30° = \frac{A}{H} = \frac{\sqrt{3}}{2} \approx 0.87 \qquad \sec 30° = \frac{H}{A} = \frac{2}{\sqrt{3}} = \frac{2\sqrt{3}}{3} \approx 1.15$$

$$\tan 30° = \frac{O}{A} = \frac{1}{\sqrt{3}} = \frac{\sqrt{3}}{3} \approx 0.58 \qquad \cot 30° = \frac{A}{O} = \frac{\sqrt{3}}{1} = \sqrt{3} \approx 1.73$$

The 30°-60°-90° triangle kills two birds with one stone because it also gives you the trig values for a 60° angle. Look at Figure 6-2 again. For the 60° angle, the $\sqrt{3}$ side of the triangle is now the *opposite* side for purposes of *SohCahToa* because it's on the opposite side of the triangle from the 60° angle. The 1-unit side becomes the *adjacent* side for the 60° angle, and the 2-unit side is still, of course, the hypotenuse. Now use *SohCahToa* again to find the trig values for the 60° angle:

$$\sin 60° = \frac{O}{H} = \frac{\sqrt{3}}{2} \approx 0.87 \qquad \csc 60° = \frac{H}{O} = \frac{2}{\sqrt{3}} = \frac{2\sqrt{3}}{3} \approx 1.15$$

$$\cos 60° = \frac{A}{H} = \frac{1}{2} \qquad \sec 60° = \frac{H}{A} = \frac{2}{1} = 2$$

$$\tan 60° = \frac{O}{A} = \frac{\sqrt{3}}{1} = \sqrt{3} \approx 1.73 \qquad \cot 60° = \frac{A}{O} = \frac{1}{\sqrt{3}} = \frac{\sqrt{3}}{3} \approx 0.58$$

The mnemonic device *SohCahToa,* along with the two oh-so-easy-to-remember right triangles in Figure 6-2, gives you the answers to 18 trig problems!

Circling the Enemy with the Unit Circle

SohCahToa only works with right triangles, and so it can only handle *acute* angles — angles less than 90°. (The angles in a triangle must add up to 180°; because a right triangle has a 90° angle, the other two angles must each be less than 90°.) With the *unit circle,* however, you can find trig values for any size angle. The *unit* circle has a radius of *one unit* and is set in an *x-y* coordinate system with its center at the origin. See Figure 6-3.

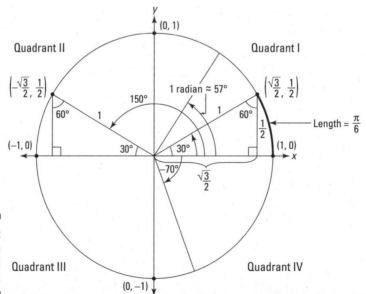

Figure 6-3: The so-called unit circle.

Figure 6-3 has quite a lot of information, but don't panic; it will all make perfect sense in a minute.

Angles in the unit circle

Measuring angles: To measure an angle in the unit circle, start at the positive *x*-axis and go *counterclockwise* to the *terminal* side of the angle.

For example, the 150° angle in Figure 6-3 begins at the positive *x*-axis and ends at the segment that hits the unit circle at $\left(-\frac{\sqrt{3}}{2}, \frac{1}{2}\right)$. If you go *clockwise* instead, you get an angle with a *negative* measure (like the −70° angle in the figure).

Measuring angles with radians

You know all about *degrees*. You know what 45° and 90° angles look like; you know that *about face* means a turn of 180° and that turning all the way around till you're back to where you started is a 360° turn.

But degrees aren't the only way to measure angles. You can also use *radians*. Degrees and radians are just two different ways to measure angles, like inches and centimeters are two ways to measure length.

Definition of *radian:* The radian measure of an angle is the length of the arc along the circumference of the unit circle cut off by the angle.

Look at the 30° angle in quadrant I of Figure 6-3. Do you see the bolded section of the circle's circumference that is cut off by that angle? Because a whole circle is 360°, that 30° angle is one-twelfth of the circle. So the length of the bold arc is one-twelfth of the circle's circumference. Circumference is given by the formula $C = 2\pi r$. This circle has a radius of 1, so its circumference equals 2π. Because the bold arc is one-twelfth of that, its length is $\frac{\pi}{6}$, which is the radian measure of the 30° angle.

360° equals 2π radians. The unit circle's circumference of 2π makes it easy to remember that 360° equals 2π radians. Half the circumference has a length of π, so 180° equals π radians.

If you focus on the fact that 180° equals π radians, other angles are easy:

- ✔ 90° is half of 180°, so 90° equals half of π, or $\frac{\pi}{2}$ radians.
- ✔ 60° is a third of 180°, so 60° equals a third of π, or $\frac{\pi}{3}$ radians.

✔ 45° is a fourth of 180°, so 45° equals a fourth of π, or $\frac{\pi}{4}$ radians.

✔ 30° is a sixth of 180°, so 30° equals a sixth of π, or $\frac{\pi}{6}$ radians.

Formulas for converting from degrees to radians and vice versa:

✔ To convert from degrees to radians, multiply the angle's measure by $\frac{\pi}{180°}$.

✔ To convert from radians to degrees, multiply the angle's measure by $\frac{180°}{\pi}$.

By the way, the word *radian* comes from *radius*. Look at Figure 6-3 again. An angle measuring 1 radian (about 57°) cuts off an arc along the circumference of this circle of the same length as the circle's radius. This is true not only of unit circles, but of circles of any size. In other words, take the radius of any circle, lay it along the circle's circumference, and that arc creates an angle of 1 radian.

Radians are preferred over degrees. In this or any other calculus book, some problems use degrees and others use radians, but radians are the preferred unit. If a problem doesn't specify the unit, do the problem in radians.

Honey, I shrunk the hypotenuse

Look at the unit circle in Figure 6-3 again. See the 30°-60°-90° triangle in quadrant I? It's the same shape but half the size of the one in Figure 6-2. Each of its sides is half as long. Because its hypotenuse now has a length of 1, and because when H is 1, $\frac{O}{H}$ equals O, the sine of the 30° angle, which equals $\frac{O}{H}$, ends up equaling the length of the opposite side. The opposite side is $\frac{1}{2}$, so that's the sine of 30°. Note that the length of the opposite side is the same as the y-coordinate of the point $\left(\frac{\sqrt{3}}{2}, \frac{1}{2}\right)$. If you figure the cosine of 30° in this triangle, it ends up equaling the length of the adjacent side, which is the same as the x-coordinate of $\left(\frac{\sqrt{3}}{2}, \frac{1}{2}\right)$. Notice that these values for sin 30° and cos 30° are the same as the ones given by the 30°-60°-90° triangle in Figure 6-2. This shows you, by the way, that shrinking a right triangle down (or blowing it up) has no effect on the trigonometric values for the angles in the triangle.

Now look at the 30°-60°-90° triangle in quadrant II in Figure 6-3. Because it's the same size as the 30°-60°-90° triangle in quadrant I, which hits the circle at $\left(\frac{\sqrt{3}}{2}, \frac{1}{2}\right)$, the triangle in quadrant II hits the circle at a point that's straight across from and symmetric to $\left(\frac{\sqrt{3}}{2}, \frac{1}{2}\right)$. The coordinates of the point in quadrant II are $\left(-\dfrac{\sqrt{3}}{2}, \dfrac{1}{2}\right)$. But remember that angles on the unit circle are all measured from the positive *x*-axis, so the hypotenuse of this triangle indicates a 150° angle; and that's the angle, not 30°, associated with the point $\left(-\dfrac{\sqrt{3}}{2}, \dfrac{1}{2}\right)$. The cosine of 150° is given by the *x*-coordinate of this point, $-\dfrac{\sqrt{3}}{2}$, and the sine of 150° equals the *y*-coordinate, $\dfrac{1}{2}$.

Coordinates on the unit circle tell you an angle's cosine and sine. The terminal side of an angle in the unit circle hits the circle at a point whose *x*-coordinate is the angle's cosine and whose *y*-coordinate is the angle's sine. Here's a mnemonic: *x* and *y* are in alphabetical order as are *cosine* and *sine*.

Putting it all together

Look at Figure 6-4. Now that you know all about the 45°-45°-90° triangle, you can easily work out — or take my word for it — that a 45°-45°-90° triangle in quadrant I hits the unit circle at $\left(\dfrac{\sqrt{2}}{2}, \dfrac{\sqrt{2}}{2}\right)$. And if you take the 30°-60°-90° triangle in quadrant I that hits the unit circle at $\left(\dfrac{\sqrt{3}}{2}, \dfrac{1}{2}\right)$ and flip it on its side, you get another 30°-60°-90° triangle with a 60° angle that hits the circle at $\left(\dfrac{1}{2}, \dfrac{\sqrt{3}}{2}\right)$. As you can see, this point has the same coordinates as those for the 30° angle but reversed.

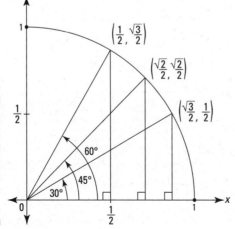

Figure 6-4: Quadrant I of the unit circle with three angles and their coordinates.

REMEMBER

How to draw a right triangle in the unit circle: Whenever you draw a right triangle in the unit circle, put the acute angle you care about at the origin — that's $(0, 0)$ — and then put the right angle on the x-axis — never on the y-axis.

TIP

$\frac{\sqrt{3}}{2}$ **is greater than** $\frac{1}{2}$. To keep from mixing up the numbers $\frac{1}{2}$ and $\frac{\sqrt{3}}{2}$ when dealing with a 30° or a 60° angle, note that because $\sqrt{3}$ is more than 1, $\frac{\sqrt{3}}{2}$ must be greater than $\frac{1}{2} \left(\frac{1}{2} = 0.5; \frac{\sqrt{3}}{2} \approx 0.87 \right)$. Thus, because a 30° angle hits the circle further out to the right than up, the x-coordinate must be greater than the y-coordinate. So, the point must be $\left(\frac{\sqrt{3}}{2}, \frac{1}{2} \right)$, not the other way around. It's vice versa for a 60° angle.

Now for the whole enchilada. Because of the symmetry in the four quadrants, the three points in quadrant I in Figure 6-4 have counterparts in the other three quadrants, giving you 12 known points. Add to these the four points on the axes, $(1, 0)$, $(0, 1)$, $(-1, 0)$, and $(0, -1)$, and you have 16 total points, each with an associated angle, as shown in Figure 6-5.

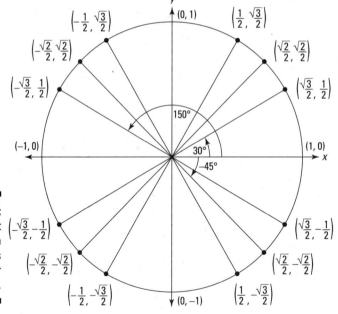

Figure 6-5:
The unit circle with 16 angles and their coordinates.

These 16 pairs of coordinates automatically give you the cosine and sine of the 16 angles. And because $\tan \theta = \frac{\sin \theta}{\cos \theta}$, you can obtain the tangent of these 16 angles by dividing an angle's *y*-coordinate by its *x*-coordinate. (Note that when the cosine of an angle equals zero, the tangent will be undefined because you can't divide by zero.) Finally, you can find the cosecant, secant, and cotangent of the 16 angles because these trig functions are just the reciprocals of sine, cosine, and tangent. (Same caution: whenever sine, cosine, or tangent equals zero, the reciprocal function will be undefined.) You've now got, at your fingertips — okay, maybe that's a bit of a stretch — the answers to 96 trig questions.

Learn the unit circle. Knowing the trig values from the unit circle is quite useful in calculus. So quiz yourself. Start by memorizing the 45°-45°-90° and the 30°-60°-90° triangles. Then picture how these triangles fit into the four quadrants of the unit circle. Use the symmetry of the quadrants as an aid. With some practice, you can get pretty quick at figuring out the values for the six trig functions of all 16 angles. (Try to do this without looking at something like Figure 6-5.) And quiz yourself with radians as well as with degrees. That would bring your total to 192 trig facts! Quick — what's the secant of 210°, and what's the cosine of $\frac{2\pi}{3}$? Here are the answers (no peeking): $-\frac{2\sqrt{3}}{3}$ and $-\frac{1}{2}$.

All Students Take Calculus. Here's a final tip to help you with the unit circle and the values of all the trig functions. Take any old unit circle (like the one in Figure 6-5) and write the initial letters of *All Students Take Calculus* in the four quadrants: Put an **A** in quadrant I, an **S** in quadrant II, a **T** in quadrant III, and a **C** in quadrant IV. These letters now tell you whether the various trig functions have positive or negative values in the different quadrants. The **A** in quadrant I tells you that **A**ll six trig functions have positive values in quadrant I. The **S** in quadrant II tells you that **S**ine (and its reciprocal, cosecant) are positive in quadrant II and that all other trig functions are negative there. The **T** in quadrant III tells you that **T**angent (and its reciprocal, cotangent) are positive in quadrant III and that the other functions are negative there. Finally, the **C** in quadrant IV tells you that **C**osine (and its reciprocal, secant) are positive there and that the other functions are negative. That's a wrap.

Graphing Sine, Cosine, and Tangent

Figure 6-6 shows the graphs of sine, cosine, and tangent, which you can, of course, produce on a graphing calculator.

Definitions of *periodic* and *period*: Sine, cosine, and tangent — and their reciprocals, cosecant, secant, and cotangent — are *periodic* functions, which means that their graphs contain a basic shape that repeats over and over indefinitely to the left and the right. The *period* of such a function is the length of one of its cycles.

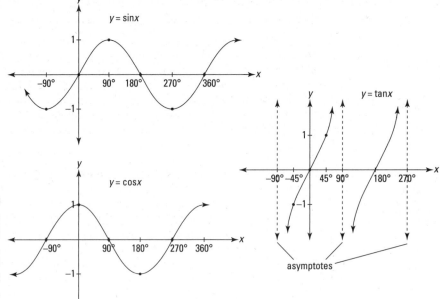

Figure 6-6:
The graphs
of the sine,
cosine, and
tangent
functions.

If you know the unit circle, you can easily reproduce these three graphs by hand. First, note that the sine and cosine graphs are the same shape — cosine is the same as sine, just slid 90° to the left. Also, notice that their simple wave shape goes as high as 1 and as low as –1 and goes on forever to the left and right, with the same shape repeating every 360°. That's the *period* of both functions, 360°. (It's no coincidence, by the way, that 360° is also once around the unit circle.) The unit circle tells you that $\sin 0° = 0$, $\sin 90° = 1$, $\sin 180° = 0$, $\sin 270° = -1$, and that $\sin 360° = 0$. If you start with these five points, you can sketch one cycle. The cycle then repeats to the left and right. You can use the unit circle in the same manner to sketch the cosine function.

Notice in Figure 6-6 that the period of the tangent function is 180°. If you remember that and the basic pattern of repeating backward *S*-shapes, sketching it isn't difficult. Because $\tan \theta = \frac{y}{x}$, you can use the unit circle to determine that $\tan(-45°) = -1$, $\tan 0° = 0$, and $\tan 45° = 1$. That gives you the points $(-45°, -1)$, $(0, 0)$, and $(45°, 1)$. Since $\tan(-90°)$ and $\tan 90°$ are both undefined (because $\frac{y}{x}$ at these points gives you a zero in the denominator), you draw *vertical asymptotes* at –90° and 90°.

Definition of *vertical asymptote*: A vertical asymptote is an imaginary line that a curve gets closer and closer to (but never touches) as the curve goes up toward infinity or down toward negative infinity. (In Chapters 7 and 8, you see more vertical asymptotes and also some horizontal asymptotes.)

The two asymptotes at $-90°$ and $90°$ and the three points at $(-45°, -1)$, $(0, 0)$, and $(45°, 1)$ show you where to sketch one backward S. The S-shapes then repeat every $180°$ to the left and the right.

Inverse Trig Functions

An inverse trig function, like any inverse function, reverses what the original function does. For example, $\sin 30° = \frac{1}{2}$, so the inverse sine function — written as $\sin^{-1}$ — reverses the input and output. Thus, $\sin^{-1}\frac{1}{2} = 30°$. It works the same for the other trig functions.

The negative 1 superscript in the sine inverse function is *not* a negative 1 power, despite the fact that it looks just like it. Raising something to the negative 1 power gives you its reciprocal, so you might think that $\sin^{-1}x$ is the reciprocal of $\sin x$, but the reciprocal of sine is cosecant, *not* sine inverse. Pretty weird that the same symbol is used to mean two different things. Go figure.

The only trick with inverse trig functions is memorizing their *ranges* — that's the interval of their outputs. Consider sine inverse, for example. Because both $\sin 30° = \frac{1}{2}$ and $\sin 150° = \frac{1}{2}$, you wouldn't know whether $\sin^{-1}\frac{1}{2}$ equals $30°$ or $150°$ unless you know how the interval of sine inverse outputs is defined. And remember, in order for something to be a function, there can't be any mystery about the output for a given input. If you reflect the sine function over the line $y = x$ to create its inverse, you get a vertical wave that isn't a function because it doesn't pass the vertical line test. (See the definition of the vertical line test in Chapter 5.) To make sine inverse a function, you have to take a small piece of the vertical wave that does pass the vertical line test. The same thing goes for the other inverse trig functions. Here are their ranges:

The range of $\sin^{-1}x$ is $\left[-\frac{\pi}{2}, \frac{\pi}{2}\right]$, or $[-90°, 90°]$.

The range of $\cos^{-1}x$ is $[0, \pi]$, or $[0°, 180°]$.

The range of $\tan^{-1}x$ is $\left[-\frac{\pi}{2}, \frac{\pi}{2}\right]$, or $[-90°, 90°]$.

The range of $\cot^{-1}x$ is $[0, \pi]$, or $[0°, 180°]$.

Note the pattern: the range of $\sin^{-1}x$ is the same as $\tan^{-1}x$, and the range of $\cos^{-1}x$ is the same as $\cot^{-1}x$.

Believe it or not, calculus authors don't agree on the ranges for the secant inverse and cosecant inverse functions. You'd think they could agree on this like they do with just about everything else in mathematics. Humph. Use the ranges given in your particular textbook. If you don't have a textbook, use the $\sin^{-1}x$ range for its cousin $\csc^{-1}x$, and use the $\cos^{-1}x$ range for $\sec^{-1}x$. (By the way, I don't refer to $\csc^{-1}x$ as the reciprocal of $\sin^{-1}x$ because it's *not* its reciprocal — even though $\csc x$ is the reciprocal of $\sin x$. Ditto for $\cos^{-1}x$ and $\sec^{-1}x$.)

Identifying with Trig Identities

Remember trig identities like $\sin^2x+\cos^2x=1$ and $\sin 2x=2\sin x\cos x$? Tell the truth now — most people remember trig identities about as well as they remember nineteenth century vice-presidents. They come in handy in calculus though, so a list of other useful ones is in the online Cheat Sheet at www.dummies.com/extras/calculus/.

Part III

Limits

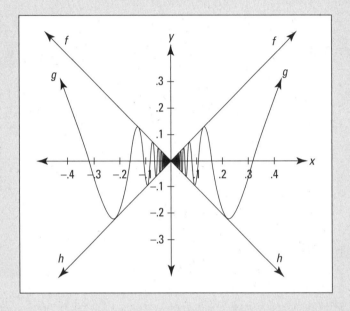

If you're up for an advanced challenge, check out my online article on the partial frac-tions technique where the denominators contain repeated linear or quadratic factors at www.dummies.com/extras/calculus.

In this part . . .

- ✔ Limits: The mathematical microscope that lets you sort of zoom in on a curve to the sub-, sub-, sub-atomic level, where it becomes straight.

- ✔ *Limits*, *asymptotes*, and *infinity*: Far out, man.

- ✔ The mathematical mumbo jumbo about *continuity*. Plus the plain English meaning: Not lifting your pencil off the paper.

- ✔ Calculating limits with algebra.

- ✔ Calculating limits with your calculator.

Chapter 7

Limits and Continuity

· ·

In This Chapter

▶ Taking a look at limits

▶ Evaluating functions with holes — break out the mothballs

▶ Exploring continuity and discontinuity

· ·

*L*imits are fundamental for both differential and integral calculus. The formal definition of a derivative involves a limit as does the definition of a definite integral. (If you're a real go-getter and can't wait to read the actual definitions, check out Chapters 9 and 13.) Now, it turns out that after you learn the shortcuts for calculating derivatives and integrals, you won't need to use the longer limit methods anymore. But understanding the mathematics of limits is nonetheless important because it forms the foundation upon which the vast architecture of calculus is built (Okay, so I got a bit carried away). In this chapter, I lay the groundwork for differentiation and integration by exploring limits and the closely related topic, continuity.

Take It to the Limit — NOT

Limits can be tricky. Don't worry if you don't grasp the concept right away.

Informal definition of *limit* (the formal definition is in a few pages): The limit of a function (if it exists) for some *x*-value *c*, is the height the function gets closer and closer to as *x* gets closer and closer to *c* from the left and the right. (***Note:*** This definition does not apply to limits where *x* approaches infinity or negative infinity. More about those limits later in the chapter and in Chapter 8.)

Got it? You're kidding! Let me say it another way. A function has a limit for a given *x*-value *c* if the function zeros in on some height as *x* gets closer and closer to the given value *c* from the left and right. Did that help? I didn't think so. It's much easier to understand limits through examples than through this sort of mumbo jumbo, so take a look at some.

Using three functions to illustrate the same limit

Consider the function $f(x) = 3x + 1$ on the left in Figure 7-1. When we say that the limit of $f(x)$ as x approaches 2 is 7, written as $\lim\limits_{x \to 2} f(x) = 7$, we mean that as x gets closer and closer to 2 from the left and the right, $f(x)$ gets closer and closer to a height of 7. By the way, as far as I know, the number 2 in this example doesn't have a formal name, but I call it the *arrow-number*. The arrow-number gives you a horizontal location in the x direction. Don't confuse it with the *answer* to the limit problem or simply the *limit*, both of which refer to a y-value or *height* of the function (7 in this example). Now, look at Table 7-1.

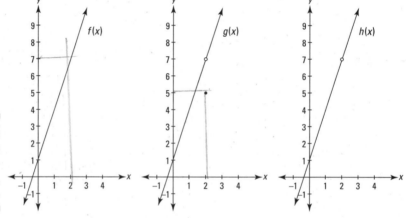

Figure 7-1:
The graphs of the functions of f, g, and h.

Table 7-1   Input and Output Values of $f(x) = 3x + 1$ as x Approaches 2

	x approaches 2 from the left					x approaches 2 from the right				
x	1	1.5	1.9	1.99	1.999	2.001	2.01	2.1	2.5	3
$f(x)$	4	5.5	6.7	6.97	6.997	7.003	7.03	7.3	8.5	10
	y approaches 7					y approaches 7				

Table 7-1 shows that y is approaching 7 as x approaches 2 from both the left and the right, and thus the limit is 7. If you're wondering what all the fuss is about — why not just plug the number 2 into x in $f(x) = 3x + 1$ and obtain the answer of 7 — I'm sure you've got a lot of company. In fact, if all functions were *continuous* (without gaps) like *f*, you *could* just plug in the arrow-number to get the answer, and this type of limit problem would basically be pointless. We need to use limits in calculus because of *discontinuous* functions like *g* and *h* that have holes.

Function *g* in the middle of Figure 7-1 is identical to *f* except for the hole at (2, 7) and the point at (2, 5). Actually, this function, *g*(*x*), would never come up in an ordinary calculus problem — I only use it to illustrate how limits work. (Keep reading. I have a bit more groundwork to lay before you see why I include it.)

The important functions for calculus are the functions like *h* on the right in Figure 7-1, which come up frequently in the study of derivatives. This third function is identical to *f*(*x*) except that the point (2, 7) has been plucked out, leaving a hole at (2, 7) and no other point where *x* equals 2.

Imagine what the table of input and output values would look like for *g*(*x*) and *h*(*x*). Can you see that the values would be identical to the values in Table 7-1 for *f*(*x*)? For both *g* and *h*, as *x* gets closer and closer to 2 from the left and the right, *y* gets closer and closer to a height of 7. For all three functions, the limit as *x* approaches 2 is 7.

This brings us to a critical point: When determining the limit of a function as *x* approaches, say, 2, the value of *f*(2) — or even whether *f*(2) exists at all — is totally irrelevant. Take a look at all three functions again where *x* = 2: *f*(2) equals 7, *g*(2) is 5, and *h*(2) doesn't exist (or, as mathematicians say, it's *undefined*). But, again, those three results are irrelevant and don't affect the answer to the limit problem.

You don't get to the limit. In a limit problem, *x* gets closer and closer to the arrow-number *c*, but technically *never gets there,* and what happens to the function when *x* equals the arrow-number *c* has *no effect* on the answer to the limit problem (though for continuous functions like *f*(*x*) the function value equals the limit answer and it can thus be used to compute the limit answer).

Sidling up to one-sided limits

One-sided limits work like regular, two-sided limits except that *x* approaches the arrow-number *c* from just the left or just the right. The most important purpose for such limits is that they're used in the formal definition of a regular limit (see the next section on the formal definition of a limit).

To indicate a one-sided limit, you put a little superscript subtraction sign on the arrow-number when x approaches the arrow-number from the left or a superscript addition sign when x approaches the arrow-number from the right. Like this:

$$\lim_{x \to 5^-} f(x) \quad \text{or} \quad \lim_{x \to 0^+} g(x)$$

Look at Figure 7-2. The answer to the regular limit problem, $\lim_{x \to 3} p(x)$, is that the limit does not exist because as x approaches 3 from the left *and* the right, $p(x)$ is not zeroing in on the same height.

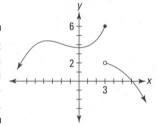

However, both one-sided limits do exist. As x approaches 3 from the left, $p(x)$ zeros in on a height of 6, and when x approaches 3 from the right, $p(x)$ zeros in on a height of 2. As with regular limits, the value of $p(3)$ has no effect on the answer to either of these one-sided limit problems. Thus,

$$\lim_{x \to 3^-} p(x) = 6 \quad \text{and} \quad \lim_{x \to 3^+} p(x) = 2$$

A function like $p(x)$ in Figure 7-2 is called a *piecewise* function because it's got separate pieces. Each part of a piecewise function has its own equation — like, for example, the following three-piece function:

$$y = \begin{cases} x^2 & \text{for} \quad x \leq 1 \\ 3x - 2 & \text{for} \quad 1 < x \leq 10 \\ x + 5 & \text{for} \quad x > 10 \end{cases}$$

Sometimes a chunk of a piecewise function connects with its neighboring chunk, in which case the function is continuous there. And sometimes, like with $p(x)$, a piece does not connect with the adjacent piece — this results in a discontinuity.

The formal definition of a limit — just what you've been waiting for

Now that you know about one-sided limits, I can give you the formal mathematical definition of a limit. Here goes:

Formal definition of limit: Let f be a function and let c be a real number.

$\lim_{x \to c} f(x)$ exists if and only if

1. $\lim_{x \to c^-} f(x)$ exists,
2. $\lim_{x \to c^+} f(x)$ exists, and
3. $\lim_{x \to c^-} f(x) = \lim_{x \to c^+} f(x)$.

Calculus books always present this as a three-part test for the existence of a limit, but condition 3 is the only one you need to worry about because 1 and 2 are built into 3. You just have to remember that you can't satisfy condition 3 if the left and right sides of the equation are both undefined or nonexistent; in other words, it is *not* true that *undefined = undefined* or that *nonexistent = nonexistent*. (I think this is why calc texts use the 3-part definition.) As long as you've got that straight, condition 3 is all you need to check.

When we say a limit exists, it means that the limit equals a *finite* number. Some limits equal infinity or negative infinity, but you nevertheless say that they *do not exist*. That may seem strange, but take my word for it. (More about infinite limits in the next section.)

Limits and vertical asymptotes

A *rational* function like $f(x) = \dfrac{(x+2)(x-5)}{(x-3)(x+1)}$ has vertical asymptotes at $x = 3$

and $x = -1$. Remember asymptotes? They're imaginary lines that the graph of a function gets closer and closer to as it goes up, down, left, or right toward infinity or negative infinity. $f(x)$ is shown in Figure 7-3.

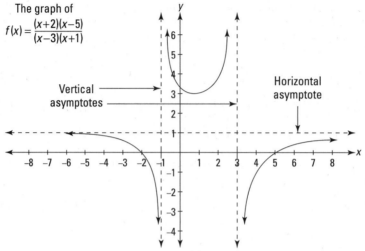

The graph of
$$f(x) = \frac{(x+2)(x-5)}{(x-3)(x+1)}$$

Vertical asymptotes

Horizontal asymptote

Figure 7-3:
A typical rational function.

Consider the limit of the function in Figure 7-3 as x approaches 3. As x approaches 3 from the left, $f(x)$ goes up to infinity, and as x approaches 3 from the right, $f(x)$ goes down to negative infinity. Sometimes it's informative to indicate this by writing,

$$\lim_{x \to 3^-} f(x) = \infty \quad \text{and} \quad \lim_{x \to 3^+} f(x) = -\infty$$

But it's also correct to say that both of these limits *do not exist* because infinity is not a real number. And if you're asked to determine the regular, two-sided limit, $\lim_{x \to 3} f(x)$, you have no choice but to say that it does not exist because the limits from the left and from the right are unequal.

Limits and horizontal asymptotes

Up till now, I've been looking at limits where x approaches a regular, finite number. But x can also approach infinity or negative infinity. Limits at infinity exist when a function has a horizontal asymptote. For example, the function in Figure 7-3 has a horizontal asymptote at $y = 1$, which the function gets closer and closer to as it goes toward infinity to the right and negative infinity to the left. (Going left, the function crosses the horizontal asymptote at $x = -7$ and then gradually comes down toward the asymptote. Going right, the function stays below the asymptote and gradually rises up toward it.) The limits equal the height of the horizontal asymptote and are written as

$$\lim_{x \to \infty} f(x) = 1 \text{ and } \lim_{x \to -\infty} f(x) = 1$$

You see more limits at infinity in Chapter 8.

Calculating instantaneous speed with limits

If you've been dozing up to now, WAKE UP! The following problem, which eventually turns out to be a limit problem, brings you to the threshold of real calculus. Say you and your calculus-loving cat are hanging out one day and you decide to drop a ball out of your second-story window. Here's the formula that tells you how far the ball has dropped after a given number of seconds (ignoring air resistance):

$h(t) = 16t^2$

(where h is the height the ball has fallen, in feet, and t is the amount of time since the ball was dropped, in seconds)

If you plug 1 into t, h is 16; so the ball falls 16 feet during the first second. During the first 2 seconds, it falls a total of $16 \cdot 2^2$, or 64 feet, and so on. Now, what if you wanted to determine the ball's speed exactly 1 second after you dropped it? You can start by whipping out this trusty ol' formula:

$$\textbf{\textit{Distance}} = \textbf{\textit{rate}} \cdot \textbf{\textit{time}}, \text{ so } \textbf{\textit{Rate}} = \frac{distance}{time}$$

Using the *rate*, or *speed* formula, you can easily figure out the ball's average speed during the 2nd second of its fall. Because it dropped 16 feet after 1 second and a total of 64 feet after 2 seconds, it fell $64 - 16$, or 48 feet from $t = 1$ second to $t = 2$ seconds. The following formula gives you the average speed:

$$\begin{aligned} Average\ speed &= \frac{total\ distance}{total\ time} \\ &= \frac{64 - 16}{2 - 1} \\ &= 48\ feet\ per\ second \end{aligned}$$

But this isn't the answer you want because the ball falls faster and faster as it drops, and you want to know its speed exactly 1 second after you drop it. The ball speeds up between 1 and 2 seconds, so this *average* speed of 48 feet per second during the 2nd second is certain to be faster than the ball's *instantaneous* speed at the end of the 1st second. For a better approximation, calculate the average speed between $t = 1$ second and $t = 1.5$ seconds. After 1.5 seconds, the ball has fallen $16 \cdot 1.5^2$, or 36 feet, so from $t = 1$ to $t = 1.5$, it falls $36 - 16$, or 20 feet. Its average speed is thus

$$Average\ speed = \frac{36 - 16}{1.5 - 1}$$
$$= 40\ \text{feet per second}$$

If you continue this process for elapsed times of a quarter of a second, a tenth of a second, then a hundredth, a thousandth, and a ten-thousandth of a second, you arrive at the list of average speeds shown in Table 7-2.

Table 7-2 **Average Speeds from 1 Second to *t* Seconds**

t seconds	2	$1\frac{1}{2}$	$1\frac{1}{4}$	$1\frac{1}{10}$	$1\frac{1}{100}$	$1\frac{1}{1,000}$	$1\frac{1}{10,000}$
Avg. speed from 1 sec. to *t* sec.	48	40	36	33.6	32.16	32.016	32.0016

As *t* gets closer and closer to 1 second, the average speeds appear to get closer and closer to 32 feet per second.

Here's the formula we used to generate the numbers in Table 7-2. It gives you the average speed between 1 second and *t* seconds:

$$Average\ speed = \frac{16t^2 - 16 \cdot 1^2}{t - 1}$$
$$= \frac{16(t^2 - 1)}{t - 1}$$
$$= \frac{16(t - 1)(t + 1)}{t - 1}$$
$$= 16t + 16 \quad (\text{where } t \neq 1)$$

(In the line immediately above, recall that *t* cannot equal 1 because that would result in a zero in the denominator of the original equation. This restriction remains in effect even after you cancel the $t - 1$.)

Figure 7-4 shows the graph of this function.

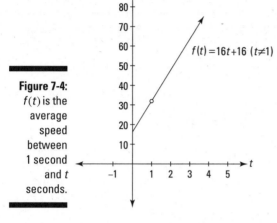

Figure 7-4:
$f(t)$ is the average speed between 1 second and t seconds.

This graph is identical to the graph of the line $y = 16t + 16$ except for the hole at $(1, 32)$. There's a hole there because if you plug 1 into t in the average speed function, you get

$$Average\ speed = \frac{16(1^2 - 1)}{1 - 1} = \frac{0}{0}$$

which is undefined. And why did you get $\frac{0}{0}$? Because you're trying to determine an average speed — which equals *total distance* divided by *elapsed time* — from $t = 1$ to $t = 1$. But from $t = 1$ to $t = 1$ is, of course, *no* time, and "during" this point in time, the ball doesn't travel any distance, so you get $\frac{zero\ feet}{zero\ seconds}$ as the average speed from $t = 1$ to $t = 1$.

Obviously, there's a problem here. Hold on to your hat, you've arrived at one of the big "Ah ha!" moments in the development of differential calculus.

Definition of *instantaneous speed*: Instantaneous speed is defined as the limit of the average speed as the elapsed time approaches zero.

For the falling-ball problem, you'd have

$$\underset{at\ t=1\ seconds}{Instantaneous\ speed} = \lim_{t \to 1} \frac{16(t^2 - 1)}{t - 1}$$

$$= \lim_{t \to 1} \frac{16(t - 1)(t + 1)}{t - 1}$$

$$= \lim_{t \to 1}(16t + 16)$$

$$= 32\ \text{feet per second}$$

The fact that the elapsed time never gets to zero doesn't affect the precision of the answer to this limit problem — the answer is exactly 32 feet per second, the height of the hole in Figure 7-4. What's remarkable about limits is that they enable you to calculate the precise, instantaneous speed at a *single* point in time by taking the limit of a function that's based on an *elapsed* time, a period between *two* points of time.

Linking Limits and Continuity

Before I expand on the material on limits from the earlier sections of this chapter, I want to introduce a related idea — *continuity.* This is such a simple concept. A *continuous* function is simply a function with no gaps — a function that you can draw without taking your pencil off the paper. Consider the four functions in Figure 7-5.

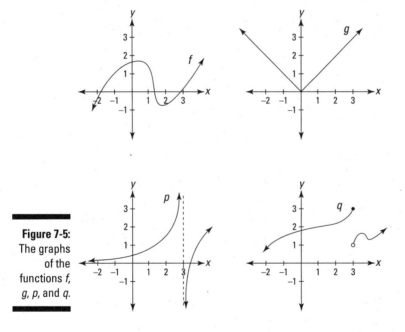

Figure 7-5:
The graphs of the functions *f*, *g*, *p*, and *q*.

Whether or not a function is continuous is almost always obvious. The first two functions in Figure 7-5, $f(x)$ and $g(x)$, have no gaps, so they're continuous. The next two, $p(x)$ and $q(x)$, have gaps at $x=3$, so they're not continuous. That's all there is to it. Well, not quite. The two functions with gaps are not continuous everywhere, but because you can draw sections of them without taking your pencil off the paper, you can say that parts of those functions are continuous. And sometimes a function is continuous everywhere it's defined. Such a function is described as being *continuous over its entire domain,* which means that its gap or gaps occur at x-values where the function is undefined. The function $p(x)$ is continuous over its entire domain; $q(x)$, on the other hand, is not continuous over its entire domain because it's not continuous at $x = 3$, which is in the function's domain. Often, the important issue is whether a function is continuous at a particular x-value. It is unless there's a gap there.

Continuity of polynomial functions: All polynomial functions are continuous everywhere.

Continuity of rational functions: All rational functions — a *rational function* is the quotient of two polynomial functions — are continuous over their entire domains. They are discontinuous at x-values not in their domains — that is, x-values where the denominator is zero.

Continuity and limits usually go hand in hand

Look at the four functions in Figure 7-5 where $x = 3$. Consider whether each function is continuous there and whether a limit exists at that x-value. The first two, f and g, have no gaps at $x = 3$, so they're continuous there. Both functions also have limits at $x = 3$, and in both cases, the limit equals the height of the function at $x = 3$, because as x gets closer and closer to 3 from the left and the right, y gets closer and closer to $f(3)$ and $g(3)$, respectively.

Functions p and q, on the other hand, are not continuous at $x = 3$ (or you can say that they're *discontinuous* there), and neither has a regular, two-sided limit at $x = 3$. For both functions, the gaps at $x = 3$ not only break the continuity, but they also cause there to be no limits there because, as you move toward $x = 3$ from the left and the right, you do not zero in on some single y-value.

So there you have it. If a function is continuous at an x-value, there must be a regular, two-sided limit for that x-value. And if there's a discontinuity at an x-value, there's no two-sided limit there . . . well, almost. Keep reading for the exception.

The hole exception tells the whole story

The hole exception is the only exception to the rule that continuity and limits go hand in hand, but it's a *huge* exception. And, I have to admit, it's a bit odd for me to say that continuity and limits *usually* go hand in hand and to talk about this *exception* because the exception is the whole point. When you come right down to it, the exception is more important than the rule. Consider the two functions in Figure 7-6.

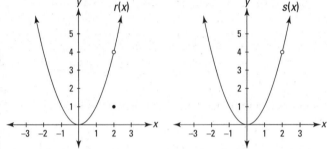

Figure 7-6:
The graphs
of the
functions r
and s.

These functions have gaps at $x = 2$ and are obviously not continuous there, but they *do* have limits as x approaches 2. In each case, the limit equals the height of the hole.

The hole exception: The only way a function can have a regular, two-sided limit where it is not continuous is where the discontinuity is an infinitesimal hole in the function.

So both functions in Figure 7-6 have the same limit as x approaches 2; the limit is 4, and the facts that $r(2) = 1$ and that $s(2)$ is undefined are irrelevant. For both functions, as x zeros in on 2 from either side, the height of the function zeros in on the height of the hole — that's the limit. This bears repeating, even an icon:

The limit at a hole: The limit at a hole is the height of the hole.

"That's great," you may be thinking. "But why should I care?" Well, stick with me for just a minute. In the falling ball example in the "Calculating instantaneous speed with limits" section earlier in this chapter, I tried to calculate the average speed during zero elapsed time. This gave me $\dfrac{\text{zero distance}}{\text{zero time}}$. Because $\dfrac{0}{0}$ is undefined, the result was a hole in the function. Function holes often come about from the impossibility of dividing zero by zero. It's these functions where the limit process is critical, and such functions are at the heart of the meaning of a derivative, and derivatives are at the heart of differential calculus.

The derivative-hole connection: A derivative always involves the undefined fraction $\dfrac{0}{0}$ and always involves the limit of a function with a hole. (If you're curious, all the limits in Chapter 9 — where the derivative is formally defined — are limits of functions with holes.)

Sorting out the mathematical mumbo jumbo of continuity

All you need to know to fully *understand* the idea of continuity is that a function is continuous at some particular *x*-value if there is no gap there. However, because you may be tested on the following formal definition, I suppose you'll want to know it.

Definition of continuity: A function $f(x)$ is *continuous* at a point $x = a$ if the following three conditions are satisfied:

1. $f(a)$ is defined,
2. $\lim\limits_{x \to a} f(x)$ exists, and
3. $f(a) = \lim\limits_{x \to a} f(x)$.

Just like with the formal definition of a limit, the definition of continuity is always presented as a 3-part test, but condition 3 is the only one you really need to worry about because conditions 1 and 2 are built into 3. You must remember, however, that condition 3 is *not* satisfied when the left and right sides of the equation are both undefined or nonexistent.

The 33333 Limit Mnemonic

Here's a great memory device that pulls a lot of information together in one swell foop. It may seem contrived or silly, but with mnemonic devices, contrived and silly work. The 33333 limit mnemonic helps you remember five groups of three things: two groups involving limits, two involving continuity, and one about derivatives. (I realize we haven't gotten to derivatives yet, but this is the best place to present this mnemonic. Take my word for it — nothing's perfect.)

First, note that the word *limit* has five letters and that there are five 3s in this mnemonic. Next, write *limit* with a lower case "l" and uncross the "t" so it becomes another "l" — like this:

l i m i l

Now, the two "l"s are for limits, the two "i"s are for continuity (notice that the letter "i" has a gap in it, thus it's not continuous), and the "m" is for slope (remember $y = mx + b$?), which is what derivatives are all about (you'll see that in Chapter 9 in just a few pages).

Each of the five letters helps you remember three things — like this:

l	i	m	i	l
3	3	3	3	3

✔ 3 parts to the definition of a limit:

Look back to the definition of a limit in "The formal definition of a limit — just what you've been waiting for" section. Remembering that it has three parts helps you remember the parts — trust me.

✔ 3 cases where a limit fails to exist:

- At a vertical asymptote — called an infinite discontinuity — like at $x = 3$ on function p in Figure 7-5.

- At a jump discontinuity, like where $x = 3$ on function q in Figure 7-5.

- With a limit at infinity of an *oscillating function* like $\sin x$ which goes up and down forever, never zeroing in on a single height.

✔ 3 parts to the definition of continuity:

Just as with the definition of a limit, remembering that the definition of continuity has 3 parts helps you remember the 3 parts (see the section "Sorting out the mathematical mumbo jumbo of continuity").

✔ 3 types of discontinuity:

- A removable discontinuity — that's a fancy term for a hole — like the holes in functions r and s in Figure 7-6.

- An infinite discontinuity like at $x = 3$ on function p in Figure 7-5.

- A jump discontinuity like at $x = 3$ on function q in Figure 7-5.

  Note that the three types of discontinuity (hole, infinite, and jump) begin with three consecutive letters of the alphabet. Since they're consecutive, there are no gaps between h, i, and j, so they're continuous letters. Hey, was this book worth the price or what?

✔ 3 cases where a derivative fails to exist:

(I explain this in Chapter 9 — keep your shirt on.)

- At any type of *discontinuity*.

- At a sharp point on a function, namely, at a *cusp* or a *corner*.

- At a *vertical tangent* (because the slope is undefined there).

Well, there you have it. Did you notice that another way this mnemonic works is that it gives you 3 cases where a limit fails to exist, 3 cases where continuity fails to exist, and 3 cases where a derivative fails to exist? *Holy triple trio of nonexistence, Batman, that's yet another 3 — the 3 topics of the mnemonic: limits, continuity, and derivatives!*

Chapter 8

Evaluating Limits

Chapter 7 introduces the concept of a limit. This chapter gets down to the nitty-gritty and presents several techniques for calculating the answers to limits problems. And while I suspect that you were radically rapt and totally transfixed by the material in Chapter 7 — and, don't get me wrong, that's important stuff — it's the problem-solving methods in this chapter that really pay the bills.

Easy Limits

A few limit problems are *very* easy. So easy that I don't have to waste your time with unnecessary introductory remarks and unneeded words that take up space and do nothing to further your knowledge of the subject — instead, I can just cut to the chase and give you only the critical facts and get to the point and get down to business and . . . Okay, so are you ready?

Limits to memorize

You should memorize the following limits. If you fail to memorize the limits in the last three bullets, you could waste *a lot* of time trying to figure them out.

   ✔ $\lim\limits_{x \to a} c = c$

     ($y = c$ is a horizontal line, so the limit — which is the function height — must equal c regardless of the arrow-number.)

   ✔ $\lim\limits_{x \to 0^+} \dfrac{1}{x} = \infty$

✔ $\lim\limits_{x\to 0^-} \dfrac{1}{x} = -\infty$

✔ $\lim\limits_{x\to \infty} \dfrac{1}{x} = 0$

✔ $\lim\limits_{x\to -\infty} \dfrac{1}{x} = 0$

✔ $\lim\limits_{x\to 0} \dfrac{\sin x}{x} = \lim\limits_{x\to 0} \dfrac{x}{\sin x} = 1$

✔ $\lim\limits_{x\to 0} \dfrac{\cos x - 1}{x} = 0$

✔ $\lim\limits_{x\to \infty} \left(1 + \dfrac{1}{x}\right)^x = e \approx 2.718$

Plugging and chugging

Plug-and-chug problems make up the second category of easy limits. Just plug the arrow-number into the limit function, and if the computation results in a number, that's your answer (but see the following warning). For example,

$$\lim_{x\to 3}(x^2 - 10) = -1$$

(Don't forget that for this method to work, the result you get after plugging in must be an ordinary number, not infinity or negative infinity or something that's undefined.)

If you're dealing with a function that's continuous everywhere (like the one in the example above) or a function that's continuous over its entire domain, this method will always work. These are *well-duh* limit problems, and, to be perfectly frank, there's really no point to them. The limit is simply the function value. If you're dealing with any other type of function, this method will only sometimes work — read on. . . .

Beware of discontinuities. The plug-and-chug method works for any type of function, including piecewise functions, *unless* there's a discontinuity at the arrow-number you plug in. In that case, if you get a number after plugging in, that number is *not* the limit; the limit might equal some other number or it might not exist. (See Chapter 7 for a description of piecewise functions.)

What happens when plugging in gives you a non-zero number over zero? If you plug the arrow-number into a limit like $\lim\limits_{x\to 5} \dfrac{10}{x-5}$ and you get any number (other than zero) divided by zero — like $\dfrac{10}{0}$ — then you know that the limit does not exist, in other words, the limit does not equal a finite number. (The answer might be infinity or negative infinity or just a plain old "does not exist.")

The "Real Deal" Limit Problems

Neither of the quick methods I present in the preceding section work for most limit problems. If you plug the arrow-number into the limit expression and the result is undefined (excluding the case covered in the previous tip), you've got a "for real" limit problem — and a bit of work to do. This is the main focus of this section. These are the interesting limit problems, the ones that likely have infinitesimal holes, and the ones that are important for differential calculus — you see more of them in Chapter 9.

When you plug in the arrow-number and the result is undefined (often because you get $\frac{0}{0}$), you can try four things: your calculator, algebra, making a limit sandwich, and L'Hôpital's rule (which is covered in Chapter 18).

Figuring a limit with your calculator

Say you want to evaluate the following limit: $\lim\limits_{x \to 5} \dfrac{x^2 - 25}{x - 5}$. The plug-and-chug method doesn't work because plugging 5 into x produces the undefined result of $\frac{0}{0}$, but as with most limit problems, you can solve this one on your calculator.

Note on calculators and other technology. With every passing year, there are more and more powerful calculators and more and more resources on the Internet that can do calculus for you. These technologies can give you an answer of, for example, $5x^3 - 4x$, when the problem calls for an algebraic answer, or an answer of, for example, $\sqrt{2}$ (not merely an approximation of 1.414), when the problem calls for a numerical answer. Older calculator models can't give you algebraic answers, and, although they can give you exact answers to many numerical problems, they can't give you an exact numerical answer like $\sqrt{2}$ — and they also can't give you an exact answer to the limit problem in the preceding paragraph.

A calculator like the TI-Nspire (or any other calculator with CAS — Computer Algebra System) can actually do that limit problem (and all sorts of much more difficult calculus problems) and give you the exact answer. The same is true of websites like Wolfram Alpha (www.wolframalpha.com).

Different calculus teachers have different policies on what technology they allow in their classes. Many do not allow the use of CAS calculators and comparable technologies because they basically do all the calculus work for you. So, the following discussion (and the rest of this book) assumes you're using a more basic calculator (like the TI-84) without CAS capability.

Method one

The first calculator method is to test the limit function with two numbers: one slightly less than the arrow-number and one slightly more than it. So here's what you do for the above problem, $\lim\limits_{x\to 5} \frac{x^2-25}{x-5}$. If you have a calculator like a Texas Instruments TI-84, enter the first number, say 4.9999, on the home screen, press the *Sto* (store) button, then the *x* button, and then the *Enter* button (this stores the number into *x*). Then enter the function, $\frac{x^2-25}{x-5}$, and hit *Enter*. The result, 9.9999, is extremely close to a round number, 10, so 10 is likely your answer. Now take a number a little more than the arrow-number, like 5.0001, and repeat the process. Since the result, 10.0001, is also very close to 10, that clinches it. The answer is 10 (almost certainly). By the way, if you're using a different calculator model, you can likely achieve the same result with the same technique or something very close to it.

Method two

The second calculator method is to produce a table of values. Enter $y = \frac{x^2-25}{x-5}$ in your calculator's graphing mode. Then go to "table set up" and enter the arrow-number, 5, as the "table start" number, and enter a small number, say 0.001, for ΔTbl — that's the size of the *x*-increments in the table. Hit the *Table* button to produce the table. Now scroll up until you can see a couple numbers less than 5, and you should see a table of values something like the one in Table 8-1.

Table 8-1	TI-84 Table for $\frac{x^2-25}{x-5}$ After Scrolling Up to 4.998	
	x	*y*
	4.998	9.998
	4.999	9.999
	5	error
	5.001	10.001
	5.002	10.002
	5.003	10.003

Because *y* gets very close to 10 as *x* zeros in on 5 from above and below, 10 is the limit (almost certainly . . . you can't be absolutely positive with these calculator methods, but they almost always work).

These calculator techniques are useful for a number of reasons. Your calculator can give you the answers to limit problems that are impossible to do algebraically. And it can solve limit problems that you could do with paper and pencil except that you're stumped. Also, for problems that you do solve on paper, you can use your calculator to check your answers. And even when you choose to solve a limit algebraically — or are required to do so — it's a good idea to create a table like Table 8-1 not just to confirm your answer, but to see how the function behaves near the arrow-number. This gives you a *numerical* grasp on the problem, which enhances your *algebraic* understanding of it. If you then look at the graph of the function on your calculator, you have a third, *graphical* or *visual* way of thinking about the problem.

Many calculus problems can be done *algebraically, graphically,* and *numerically.* When possible, use two or three of the approaches. Each approach gives you a different perspective on a problem and enhances your grasp of the relevant concepts.

Use the calculator methods to supplement algebraic methods, but don't rely too much on them. First of all, the non-CAS-calculator techniques won't allow you to deduce an exact answer unless the numbers your calculator gives you are getting close to a number you recognize — like 9.999 is close to 10, or 0.333332 is close to $\frac{1}{3}$; or perhaps you recognize that 1.414211 is very close to $\sqrt{2}$. But if the answer to a limit problem is something like $\frac{1}{2\sqrt{3}}$, you probably won't recognize it. The number $\frac{1}{2\sqrt{3}}$ is approximately equal to 0.288675. When you see numbers in your table close to that decimal, you won't recognize $\frac{1}{2\sqrt{3}}$ as the limit — unless you're an Archimedes, a Gauss, or a Ramanujan (members of the mathematics hall of fame). However, even when you don't recognize the *exact* answer in such cases, you can still learn an approximate answer, in decimal form, to the limit question.

Gnarly functions may stump your calculator. The second calculator limitation is that it won't work at all with some peculiar functions like $\lim\limits_{x \to 5} \sqrt[25]{x-5} \cdot \sin\left(\frac{1}{x-5}\right)$. This limit equals zero, but you can't get that result with your calculator.

By the way, even when the non-CAS-calculator methods work, these calculators can do some quirky things from time to time. For example, if you're solving a limit problem where *x* approaches 3, and you put numbers in your calculator that are *too* close to 3 (like 3.0000000001), you can get too close to the calculator's maximum decimal length. This can result in answers that get *further* from the limit answer, even as you input numbers closer and closer to the arrow-number.

The moral of the story is that you should think of your calculator as one of several tools at your disposal for solving limits — not as a substitute for algebraic techniques.

Solving limit problems with algebra

You use two main algebraic techniques for "real" limit problems: factoring and conjugate multiplication. I lump other algebra techniques in the section "Miscellaneous algebra." All algebraic methods involve the same basic idea. When substitution doesn't work in the original function — usually because of a hole in the function — you can use algebra to manipulate the function until substitution does work (it works because your manipulation plugs the hole).

Fun with factoring

Here's an example. Evaluate $\lim\limits_{x \to 5} \dfrac{x^2 - 25}{x - 5}$, the same problem you did with a calculator in the preceding section:

1. **Try plugging 5 into x — you should *always* try substitution first.**

   You get $\dfrac{0}{0}$ — no good, on to plan B.

2. **$x^2 - 25$ can be factored, so do it.**

   $$\lim\limits_{x \to 5} \frac{x^2 - 25}{x - 5} = \lim\limits_{x \to 5} \frac{(x - 5)(x + 5)}{x - 5}$$

3. **Cancel the $(x - 5)$ from the numerator and denominator.**

   $$= \lim\limits_{x \to 5}(x + 5)$$

4. **Now substitution will work.**

   $$= 5 + 5 = 10$$

So, $\lim\limits_{x \to 5} \dfrac{x^2 - 25}{x - 5} = 10$, confirming the calculator answer.

By the way, the function you got after canceling the $(x - 5)$, namely $y = (x + 5)$, is identical to the original function, $y = \dfrac{x^2 - 25}{x - 5}$, except that the hole in the original function at $(5, 10)$ has been plugged. And note that the limit as x approaches 5 is 10, which is the height of the hole at $(5, 10)$.

Conjugate multiplication — no, this has nothing to do with procreation

Try this method for fraction functions that contain square roots. Conjugate multiplication *rationalizes* the numerator or denominator of a fraction, which means getting rid of square roots. Try this one: Evaluate $\lim\limits_{x \to 4} \dfrac{\sqrt{x} - 2}{x - 4}$.

1. **Try substitution.**

   Plug in 4: that gives you $\frac{0}{0}$ — time for plan B.

2. **Multiply the numerator *and* denominator by the conjugate of $\sqrt{x}-2$, which is $\sqrt{x}+2$.**

   Definition of *conjugate*: The conjugate of a two-term expression is just the same expression with subtraction switched to addition or vice versa. The product of conjugates always equals the first term squared minus the second term squared.

   Now do the rationalizing.

   $$\lim_{x \to 4} \frac{\sqrt{x}-2}{x-4}$$

   $$=\lim_{x \to 4} \frac{(\sqrt{x}-2)}{(x-4)} \cdot \frac{(\sqrt{x}+2)}{(\sqrt{x}+2)}$$

   $$=\lim_{x \to 4} \frac{\left(\sqrt{x}\right)^2 - 2^2}{(x-4)(\sqrt{x}+2)}$$

   $$=\lim_{x \to 4} \frac{(x-4)}{(x-4)(\sqrt{x}+2)}$$

3. **Cancel the $(x-4)$ from the numerator and denominator.**

   $$=\lim_{x \to 4} \frac{1}{\sqrt{x}+2}$$

4. **Now substitution works.**

   $$=\frac{1}{\sqrt{4}+2}=\frac{1}{4}$$

So, $\lim_{x \to 4} \dfrac{\sqrt{x}-2}{x-4} = \dfrac{1}{4}$.

As with the factoring example, this rationalizing process plugged the hole in the original function. In this example, 4 is the arrow-number, $\frac{1}{4}$ is the limit answer, and the function $\dfrac{\sqrt{x}-2}{x-4}$ has a hole at $(4, \frac{1}{4})$.

Miscellaneous algebra

When factoring and conjugate multiplication don't work, try some other basic algebra, like adding or subtracting fractions, multiplying or dividing fractions, canceling, or some other form of simplification. Here's an example:

Evaluate $\lim\limits_{x \to 0} \dfrac{\dfrac{1}{x+4} - \dfrac{1}{4}}{x}$.

1. **Try substitution.**

   Plug in 0: That gives you $\frac{0}{0}$ — no good.

2. **Simplify the complex fraction (that's a big fraction that contains little fractions) by multiplying the numerator and denominator by the least common denominator of the little fractions, namely $4(x + 4)$.**

   Note: You can also simplify a complex fraction by adding or subtracting the little fractions in the numerator and/or denominator, but the method described here is a bit quicker.

   $$\lim_{x \to 0} \frac{\frac{1}{x+4} - \frac{1}{4}}{x}$$

   $$= \lim_{x \to 0} \frac{\left(\frac{1}{x+4} - \frac{1}{4}\right)}{x} \cdot \frac{4(x+4)}{4(x+4)}$$

   $$= \lim_{x \to 0} \frac{4-(x+4)}{4x(x+4)}$$

   $$= \lim_{x \to 0} \frac{-x}{4x(x+4)}$$

   $$= \lim_{x \to 0} \frac{-1}{4(x+4)}$$

3. **Now substitution works.**

   $$= \frac{-1}{4(0+4)} = -\frac{1}{16}$$

That's the limit.

Take a break and make yourself a limit sandwich

When algebra doesn't work, try making a limit sandwich. The best way to understand the *sandwich* or *squeeze* method is by looking at a graph. See Figure 8-1.

Look at functions f, g, and h in Figure 8-1: g is sandwiched between f and h. Since near the arrow-number of 2, f is always higher than or the same height as g, and g is always higher than or the same height as h, and since $\lim_{x \to 2} f(x) = \lim_{x \to 2} h(x)$, then $g(x)$ must have the same limit as x approaches 2 because it's sandwiched or squeezed between f and h. The limit of both f and h as x approaches 2 is 3. So, 3 has to be the limit of g as well. It's got nowhere else to go.

Figure 8-1:
The sandwich method for solving a limit. Functions f and h are the bread, and g is the salami.

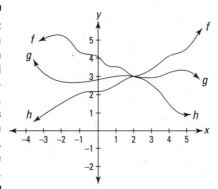

Here's another example: Evaluate $\lim\limits_{x \to 0} \left(x \sin \dfrac{1}{x} \right)$.

1. **Try substitution.**

   Plug 0 into x. That gives you $0 \cdot \sin \dfrac{1}{0}$ — no good, can't divide by zero. On to plan B.

2. **Try the algebraic methods or any other tricks you have up your sleeve.**

   Knock yourself out. You can't do it. Plan C.

3. **Try your calculator.**

   It's always a good idea to see what your calculator tells you even if this is a "show your work" problem. To graph this function, set your graphing calculator's mode to *radian* and the window to

$$x \min = -0.4$$
$$x \max = 0.4$$
$$y \min = -0.3$$
$$y \max = 0.3$$

   Figure 8-2 shows what the graph looks like.

   It definitely looks like the limit of g is zero as x approaches zero from the left and the right. Now, check the table of values on your calculator (set *TblStart* to 0 and ΔTbl to 0.001). Table 8-2 gives some of the values from the calculator table.

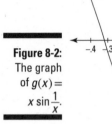

Figure 8-2:
The graph
of $g(x) = $
$x \sin\dfrac{1}{x}$.

Table 8-2	Table of Values for $g(x) = x \sin\dfrac{1}{x}$
x	**$g(x)$**
0	Error
0.001	0.0008269
0.002	−0.000936
0.003	0.0009565
0.004	−0.003882
0.005	−0.004366
0.006	−0.000969
0.007	−0.006975
0.008	−0.004928
0.009	−0.008234

These numbers sort of look like they're getting closer and closer to zero as x gets close to zero, but they're not convincing. This type of table doesn't work so great for oscillating functions like sine or cosine. (Some function values on the table, for example −0.000969 for $x = 0.006$, are closer to zero than other values higher on the table where x is smaller. That's the opposite of what we want to see.)

A better way of seeing that the limit of g is zero as x approaches zero is to use the first calculator method I discuss in the section "Figuring a limit with your calculator." Enter the function on the home screen and successively plug in the x-values listed in Table 8-3 to obtain the corresponding function values. (*Note:* Don't be confused: Table 8-3 is called a "table," but it is not a table generated by a calculator's table function. Get it?)

Table 8-3	Another Table of Values for $g(x) = x \sin\dfrac{1}{x}$	
x	**g (x)**	
0.1	−0.054	
0.01	−0.0051	
0.001	0.00083	
0.0001	−0.000031	
0.00001	0.00000036	

Now you can definitely see that g is headed toward zero.

4. **Now you need to prove the limit mathematically even though you've already solved it on your calculator. To do this, make a limit sandwich. (Fooled you — bet you thought Step 3 was the last step.)**

The hard part about using the sandwich method is coming up with the "bread" functions. (Functions f and h are the bread, and g is the salami.) There's no automatic way of doing this. You've got to think about the shape of the salami function, and then use your knowledge of functions and your imagination to come up with some good prospects for the bread functions.

Because the range of the sine function is from negative 1 to positive 1, whenever you multiply a number by the sine of anything, the result either stays the same distance from zero or gets closer to zero. Thus, $x \sin\dfrac{1}{x}$ will never get above $|x|$ or below $-|x|$. So try graphing the functions $f(x) = |x|$ and $h(x) = -|x|$ along with $g(x)$ to see if f and h make adequate bread functions for g. Figure 8-3 shows that they do.

We've shown — though perhaps not to a mathematician's satisfaction, *egad!* — that $f(x) \geq g(x) \geq h(x)$. And because $\lim\limits_{x \to 0} f(x) = \lim\limits_{x \to 0} h(x) = 0$, it follows that $g(x)$ must have the same limit: voilà — $\lim\limits_{x \to 0} g(x) = 0$.

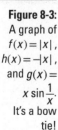

Figure 8-3: A graph of $f(x) = |x|$, $h(x) = -|x|$, and $g(x) = x \sin\dfrac{1}{x}$. It's a bow tie!

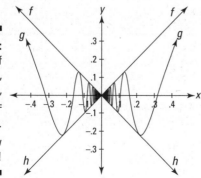

The long and winding road

Consider the function, $g(x) = x \sin \frac{1}{x}$, shown in Figures 8-2 and 8-3 and discussed in the section about making a limit sandwich. It's defined everywhere except at zero. If we now alter it slightly — by renaming it $f(x)$ and then defining $f(0)$ to be 0 — we create a new function with bizarre properties. The function is now continuous everywhere; in other words, it has no gaps. But at $(0,0)$, it seems to contradict the basic idea of continuity that says you can trace the function without taking your pencil off the paper.

Imagine starting anywhere on $f(x)$ — which looks exactly like $g(x)$ in Figures 8-2 and 8-3 — to the left of the y-axis and driving along the winding road toward the origin, $(0,0)$. Get this: You can start your drive as close to the origin as you like — how about the width of a proton away from $(0,0)$ — and the length of road between you and $(0,0)$ is *infinitely* long! That's right. It winds up and down with such increasing frequency as you get closer and closer to $(0,0)$ that the length of your drive is actually infinite, despite the fact that each "straight-away" is getting shorter and shorter. On this long and winding road, you'll never get to her door.

This altered function is clearly continuous at every point — with the possible exception of $(0,0)$ — because it's a smooth, connected, winding road. And because $\lim\limits_{x \to 0} x \sin \frac{1}{x} = 0$ (see the limit sandwich section for proof), and because $f(0)$ is defined to be 0, the three-part test for continuity at 0 is satisfied. The function is thus continuous everywhere.

But tell me, how can the curve ever reach $(0,0)$ or connect to $(0,0)$ from the left (or the right)? Assuming you can traverse an infinite distance by driving infinitely fast, when you finally drive through the origin, are you on one of the up legs of the road or one of the down legs? Neither seems possible because no matter how close you are to the origin, you have an infinite number of legs and an infinite number of turns ahead of you. There is no last turn before you reach $(0,0)$. So it seems that the function can't connect to the origin and that, therefore, it can't be continuous there — despite the fact that the math tells us that it is.

Here's another way of looking at it. Imagine a vertical line drawn on top of the function at $x = -0.2$. Now, keeping the line vertical, slowly slide the line to the right over the function until you pass over $(0,0)$. There are no gaps in the function, so at every instance, the vertical line crosses the function somewhere. Think about the point where the line intersects with the function. As you drag the line to the right, that point travels along the function, winding up and down along the road, and, as you drag the line over the origin, the point reaches and then passes $(0,0)$. Now tell me this: When the point hits $(0,0)$, is it on its way up or down? How can you reconcile all this? I wish I knew.

Stuff like this really messes with your mind.

Evaluating Limits at Infinity

In the previous sections, I look at limits as x approaches a finite number, but you can also have limits where x approaches infinity or negative infinity. Consider the function $f(x) = \dfrac{1}{x}$ and check out its graph in Figure 8-4.

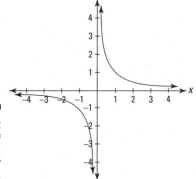

Figure 8-4:
The graph of
$f(x) = \dfrac{1}{x}$.

You can see on the graph (in the first quadrant) that as x gets bigger and bigger — in other words, as x approaches infinity — the height of the function gets lower and lower but never gets to zero. This is confirmed by considering what happens when you plug bigger and bigger numbers into $\dfrac{1}{x}$: the outputs get smaller and smaller and approach zero. This graph thus has a horizontal asymptote of $y = 0$ (the x-axis), and we say that $\lim\limits_{x \to \infty} \dfrac{1}{x} = 0$. The fact that x never actually reaches infinity and that f never gets to zero has no relevance. When we say that $\lim\limits_{x \to \infty} \dfrac{1}{x} = 0$, we mean that as x gets bigger and bigger without end, f is closing in on a height of zero (or f is ultimately getting infinitely close to a height of zero). If you look at the third quadrant, you can see that the function f also approaches zero as x approaches negative infinity, which is written as $\lim\limits_{x \to -\infty} \dfrac{1}{x} = 0$.

Limits at infinity and horizontal asymptotes

Limits at infinity and horizontal asymptotes go hand in hand. Determining the limit of a function as x approaches infinity or negative infinity is the same as finding the height of the horizontal asymptote.

Let's begin by considering rational functions. Here's how you find the limit at infinity and negative infinity (and the height of the horizontal asymptote) of a rational function (something like $f(x) = \frac{3x-7}{2x+8}$). First, note the degree of the numerator (that's the highest power of x in the numerator) and the degree of the denominator. There are three cases:

- ✔ If the degree of the numerator is greater than the degree of the denominator, for example $f(x) = \frac{6x^4 + x^3 - 7}{2x^2 + 8}$, there's no horizontal asymptote, and the limit of the function as x approaches infinity (or negative infinity) does not exist (the limit will be positive or negative infinity).

- ✔ If the degree of the denominator is greater than the degree of the numerator, for example $g(x) = \frac{4x^2 - 9}{x^3 + 12}$, the x-axis (that's the line $y=0$) is the horizontal asymptote, and $\lim\limits_{x \to \infty} g(x) = \lim\limits_{x \to -\infty} g(x) = 0$.

- ✔ If the degrees of the numerator and denominator are equal, take the coefficient of the highest power of x in the numerator and divide it by the coefficient of the highest power of x in the denominator. That quotient gives you the answer to the limit problem and the height of the asymptote. For example, if $h(x) = \frac{4x^3 - 10x + 1}{5x^3 + 2x^2 - x}$, $\lim\limits_{x \to \infty} h(x) = \lim\limits_{x \to -\infty} h(x) = \frac{4}{5}$, and h has a horizontal asymptote at $y = \frac{4}{5}$.

Talk like a professor. To impress your friends, point your index finger upward, raise one eyebrow, and say in a professorial tone, "In a rational function where the numerator and denominator are of equal degrees, the limit of the function as x approaches infinity or negative infinity equals the quotient of the coefficients of the leading terms. A horizontal asymptote occurs at this same value."

$\frac{\infty}{\infty}$ **does *not* equal 1.** Substitution doesn't work for the problems in this section. If you try plugging ∞ into x in any of the rational functions in this section, you get $\frac{\infty}{\infty}$ but that does *not* necessarily equal 1 ($\frac{\infty}{\infty}$ sometimes equals 1, but it often does not). A result of $\frac{\infty}{\infty}$ tells you nothing about the answer to a limit problem.

Solving limits at infinity with a calculator

Here's a problem that can't be done by the method in the previous section because it's not a rational function: $\lim\limits_{x \to \infty} \left(\sqrt{x^2 + x} - x \right)$. But it's a snap with a calculator. Enter the function in graphing mode, then go to *table setup* and set *TblStart* to 100,000 and ΔTbl to 100,000. Table 8-4 shows the results.

Table 8-4	Table of Values for $\sqrt{x^2+x}-x$
x	y
100,000	0.4999988
200,000	0.4999994
300,000	0.4999996
400,000	0.4999997
500,000	0.4999998
600,000	0.4999998
700,000	0.4999998
800,000	0.4999998
900,000	0.4999999

You can see that y is getting extremely close to 0.5 as x gets larger and larger. So 0.5 is the limit of the function as x approaches infinity, and there's a horizontal asymptote at $y = 0.5$. If you have any doubts that the limit equals 0.5, go back to *table setup* and put in a humongous *TblStart* and ΔTbl, say 1,000,000,000, and check the table results again. All you see is a column of 0.5s. That's the limit. (By the way, unlike with the rational functions in the two previous sections, the limit of this function as x approaches negative infinity doesn't equal the limit as x approaches positive infinity:

$$\lim_{x\to-\infty}\left(\sqrt{x^2+x}-x\right)=\infty,$$ because when you plug in $-\infty$ you get $\infty+\infty$ which equals ∞.) One more thing: Just as with regular limits, using a non-CAS (Computer Algebra System) calculator for infinite limits won't give you an exact answer unless the numbers in the table are getting close to a number you recognize, like 0.5.

$\infty - \infty$ **does *not* equal zero.** Substitution does not work for the problem above, $\lim_{x\to\infty}\left(\sqrt{x^2+x}-x\right)$. If you plug ∞ into x, you get $\infty - \infty$ which does *not* necessarily equal zero ($\infty - \infty$ sometimes equals zero, but it often does not). A result of $\infty - \infty$ tells you nothing about the answer to a limit problem.

Solving limits at infinity with algebra

Now try some algebra for the problem $\lim_{x\to\infty}\left(\sqrt{x^2+x}-x\right)$. You got the answer with a calculator, but all things being equal, it's better to solve the problem algebraically because then you have a mathematically airtight answer. The calculator answer in this case is *very* convincing, but it's not mathematically rigorous, so if you stop there, the math police may get you.

1. **Try substitution — always a good idea.**

   No good. You get $\infty - \infty$, which tells you nothing — see the Warning in the previous section. On to plan B.

   Because $\left(\sqrt{x^2 + x} - x \right)$ contains a square root, the conjugate multiplication method would be a natural choice, except that that method is used for fraction functions. Well, just put $\left(\sqrt{x^2 + x} - x \right)$ over the number 1 and, voilà, you've got a fraction: $\dfrac{\sqrt{x^2 + x} - x}{1}$. Now do the conjugate multiplication.

2. **Multiply the numerator and denominator by the conjugate of $\left(\sqrt{x^2 + x} - x \right)$ and simplify.**

$$\lim_{x \to \infty} \frac{\sqrt{x^2 + x} - x}{1}$$

$$= \lim_{x \to \infty} \frac{(\sqrt{x^2 + x} - x)}{1} \cdot \frac{(\sqrt{x^2 + x} + x)}{(\sqrt{x^2 + x} + x)}$$

$$= \lim_{x \to \infty} \frac{x^2 + x - x^2}{\sqrt{x^2 + x} + x} \qquad \begin{array}{l} \text{(First, cancel the } x^2\text{s in the numerator.} \\ \text{Then factor } x \text{ out of the denominator.} \\ \text{Yes, you heard that right.)} \end{array}$$

$$= \lim_{x \to \infty} \frac{x}{x\left(\sqrt{1 + \frac{1}{x}} + 1 \right)} \qquad \text{(Now, cancel the } x\text{s.)}$$

$$= \lim_{x \to \infty} \frac{1}{\sqrt{1 + \frac{1}{x}} + 1}$$

3. **Now substitution works.**

$$= \frac{1}{\sqrt{1 + \frac{1}{\infty}} + 1}$$

$$= \frac{1}{\sqrt{1 + 0} + 1} \qquad \text{(Recall that } \lim_{x \to \infty} \frac{1}{x} = 0 \text{ from the ``Limits to memorize'' section.)}$$

$$= \frac{1}{1 + 1} = \frac{1}{2}$$

Thus, $\lim\limits_{x \to \infty} \left(\sqrt{x^2 + x} - x \right) = \frac{1}{2}$, which confirms the calculator answer.

Part IV
Differentiation

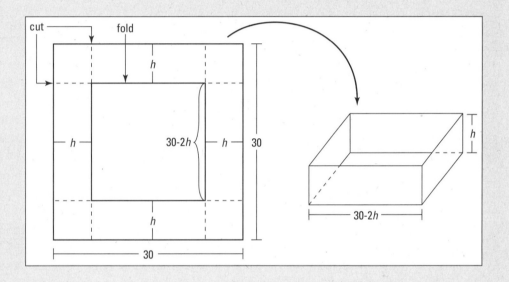

In this part . . .

- ✔ The meaning of a derivative: It's a slope and a rate — more specifically, a derivative tells you how fast y is changing compared to x.

- ✔ How to calculate derivatives with the product rule, the quotient rule, and the chain rule.

- ✔ Implicit differentiation, logarithmic differentiation, and the differentiation of inverse functions.

- ✔ What a derivative tells you about the shape of a curve: Local minimums, local maximums, steepness, inflection points, concavity, critical numbers, and so on.

- ✔ Differentiation word problems: Position, velocity, and acceleration, optimization, related rates, linear approximation, and tangent and normal lines.

Chapter 9

Differentiation Orientation

• •

In This Chapter

▶ Discovering the simple algebra behind the calculus

▶ Getting a grip on weird calculus symbols

▶ Differentiating with Laurel and Hardy

▶ Finding the derivatives of lines and curves

▶ Tackling the tangent line problem and the difference quotient

• •

Differential calculus is the mathematics of *change* and the mathematics of *infinitesimals*. You might say that it's the mathematics of infinitesimal changes — changes that occur every gazillionth of a second.

Without differential calculus — if you've got only algebra, geometry, and trigonometry — you're limited to the mathematics of things that either don't change or that change or move at an *unchanging* rate. Remember those problems from algebra? One train leaves the station at 3 p.m. going west at 80 mph. Two hours later another train leaves going east at 50 mph . . . . You can handle such a problem with algebra because the speeds or rates are unchanging. Our world, however, isn't one of unchanging rates — rates are in constant flux.

Think about putting man on the moon. Apollo 11 took off from a *moving* launch pad (the earth is both rotating on its axis and revolving around the sun). As the Apollo flew higher and higher, the friction caused by the atmosphere and the effect of the earth's gravity were changing not just every second, not just every millionth of a second, but every *infinitesimal* fraction of a second. The spacecraft's weight was also constantly changing as it burned fuel. All of these things influenced the rocket's changing speed. On top of all that, the rocket had to hit a *moving* target, the moon. All of these things were changing, and their rates of change were changing. Say the rocket was going 1,000 mph one second and 1,020 mph a second later — during that one second, the rocket's speed literally passed through the *infinite* number of different speeds between 1,000 and 1,020 mph. How can you do the math for these ephemeral things that change every *infinitesimal* part of a second? You can't do it without differential calculus.

And differential calculus is used for all sorts of terrestrial things as well. Much of modern economic theory, for example, relies on differentiation. In economics, everything is in constant flux. Prices go up and down, supply and demand fluctuate, and inflation is constantly changing. These things are constantly changing, and the ways they affect each other are constantly changing. You need calculus for this.

Differential calculus is one of the most practical and powerful inventions in the history of mathematics. So let's get started already.

Differentiating: It's Just Finding the Slope

Differentiation is the first of the two major ideas in calculus (the other is integration, which I cover in Part V). Differentiation is the process of finding the derivative of a function like $y = x^2$. The *derivative* is just a fancy calculus term for a simple idea you know from algebra: slope. *Slope,* as you know, is the fancy algebra term for steepness. And *steepness* is the fancy word for . . . No! Steepness is the *ordinary* word you've known since you were a kid, as in, "Hey, this road sure is steep." Everything you study in differential calculus all relates back to the simple idea of steepness.

In *differential* calculus, you study *differentiation,* which is the process of *deriving* — that's finding — *derivatives.* These are big words for a simple idea: Finding the *steepness* or *slope* of a line or curve. Throw some of these terms around to impress your friends. By the way, the root of the words *differential* and *differentiation* is *difference* — I explain the connection at the end of this chapter in the section on the *difference quotient.*

Consider Figure 9-1. A steepness of $\frac{1}{2}$ means that as the stickman walks one foot to the right, he goes up $\frac{1}{2}$ foot; where the steepness is 3, he goes up 3 feet as he walks 1 foot to the right. Where the steepness is zero, he's at the top, going neither up nor down; and where the steepness is negative, he's going down. A steepness of –2, for example, means he goes *down* 2 feet for every foot he goes to the right. This is shown more precisely in Figure 9-2.

Negative slope: To remember that going down to the right (or up to the left) is a *negative* slope, picture an uppercase *N,* as shown in Figure 9-3.

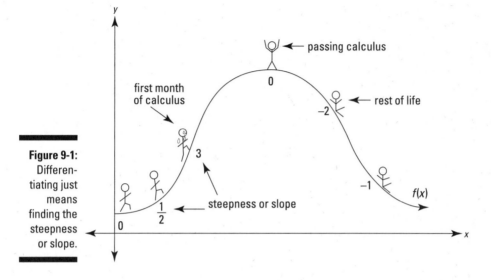

passing calculus

first month
of calculus

rest of life

0

−2

3

−1

$f(x)$

$\frac{1}{2}$

steepness or slope

0

Figure 9-1:
Differen-
tiating just
means
finding the
steepness
or slope.

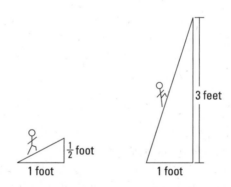

3 feet

$\frac{1}{2}$ foot

1 foot

1 foot

English:	steepness = $\frac{1}{2}$	steepness = 3
Algebra:	slope = $\frac{1}{2}$	slope = 3
Calculus:	$\frac{dy}{dx} = \frac{1}{2}$	$\frac{dy}{dx} = 3$

Figure 9-2:
The
derivative =
slope =
steepness.

($\frac{dy}{dx}$, pronounced *dee y dee x*, is one of the many
symbols for the derivative — see sidebar.)

Figure 9-3:
This *N*
line has a
Negative
slope.

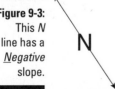

N

Don't be among the legions of students who mix up the slopes of vertical and horizontal lines. How steep is a flat, horizontal road? Not steep at all, of course. Zero steepness. So, a horizontal line has a slope of *zero*. (Like where the stick man is at the top of the hill in Figure 9-1.) What's it like to drive up a vertical road? You can't do it. And you can't get the slope of a vertical line — it doesn't exist, or, as mathematicians say, it's *undefined*.

Variety is the spice of life

Everyone knows that $3^2 = 9$. Now, wouldn't it be weird if the next time you read this math fact, it was written as $^2 3 = 9$ or $_2 3 = 9$? How does $\overset{2}{3} = 9$ grab you? Or $\underset{2}{3} = 9$? Variety is *not* the spice of mathematics. When mathematicians decide on a way of expressing an idea, they stick to it — except, that is, with calculus. Are you ready? Hold on to your hat. All of the following are different symbols for the derivative — they all mean *exactly* the same thing: $\frac{dy}{dx}$ or $\frac{df}{dx}$ or $\frac{df(x)}{dx}$ or $D_x y$ or $\frac{d}{dx} f(x)$ or $f'(x)$ or y' or $\dot{f}$ or $\dot{y}$ or $D_x f$ or Df or $D_x f(x)$. There are more. Now, you've got two alternatives: 1) Beat your head against the wall trying to figure out things like why some author uses one symbol one time and a different symbol another time, and what exactly does the d or f mean anyway, and so on and so on, or 2) Don't try to figure it out; just treat these different symbols like words in different languages for the same idea — in other words, *don't sweat it.* I strongly recommend the second option.

The slope of a line

Keep going with the slope idea — by now you should know that slope is what differentiation is all about. Take a look at the graph of the line, $y = 2x + 3$, in Figure 9-4.

You remember from algebra — I'm *totally confident* of this — that you can find points on this line by plugging numbers into x and calculating y: plug 1 into x and y equals 5, which gives you the point located at $(1, 5)$; plug 4 into x and y equals 11, giving you the point $(4, 11)$, and so on.

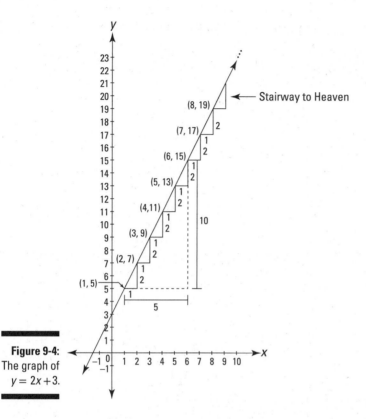

Stairway to Heaven

Figure 9-4:
The graph of
$y = 2x + 3.$

I'm sure you also remember how to calculate the slope of this line. I realize that no calculation is necessary here — you go up 2 as you go over 1, so the slope is automatically 2. You can also simply note that $y = 2x + 3$ is in slope-intercept form ($y = mx + b$) and that, since $m = 2$, the slope is 2. (See Chapter 5 if you want to review $y = mx + b$.) But bear with me because you need to know what follows. First, recall that

$$Slope = \frac{rise}{run} = \frac{y_2 - y_1}{x_2 - x_1}$$

The *rise* is the distance you go up (the vertical part of a stair step), and the *run* is the distance you go across (the horizontal part of a stair step). Now, take any two points on the line, say, (1, 5) and (6, 15), and figure the rise and the run. You *rise* up 10 from (1, 5) to (6, 15) because 5 plus 10 is 15 (or you could say that 15 minus 5 is 10). And you *run* across 5 from (1, 5) to (6, 15) because 1 plus 5 is 6 (or in other words, 6 minus 1 is 5). Next, you divide to get the slope:

$$Slope = \frac{rise}{run} = \frac{10}{5} = 2$$

Here's how you do the same problem using the slope formula:

$$Slope = \frac{y_2 - y_1}{x_2 - x_1}$$

Plug in the points $(1, 5)$ and $(6, 15)$:

$$Slope = \frac{15 - 5}{6 - 1} = \frac{10}{5} = 2$$

Okay, let's summarize what we know about this line. Table 9-1 shows six points on the line and the unchanging slope of 2.

Table 9-1	Points on the Line $y = 2x + 3$ and the Slope at Those Points						
x (horizontal position)	1	2	3	4	5	6	etc.
y (height)	5	7	9	11	13	15	etc.
slope	2	2	2	2	2	2	etc.

The derivative of a line

The preceding section showed you the algebra of slope. Now, here's the calculus. The derivative (the slope) of the line in Figure 9-4 is always 2, so you write

$$\frac{dy}{dx} = 2 \quad \text{(Read } dee\ y\ dee\ x\ equals\ 2.)$$

Another common way of writing the same thing is

$$y' = 2 \quad \text{(Read } y\ prime\ equals\ 2.)$$

And you say,

The derivative of the function, $y = 2x + 3$, is 2.

(Read *The derivative of the function, $y = 2x + 3$, is 2.* That was a joke.)

The Derivative: It's Just a Rate

Here's another way to understand the idea of a derivative that's even more fundamental than the concept of slope: A derivative is a *rate*. So why did I start the chapter with *slope?* Because slope is in some respects the easier of the two concepts, and slope is the idea you return to again and again in this book and any other calculus textbook as you look at the graphs of dozens and dozens of functions. But before you've got a slope, you've got a rate. A slope is, in a sense, a picture of a rate; the rate comes first, the picture of it comes second. Just like you can have a function before you see its graph, you can have a rate before you see it as a slope.

Calculus on the playground

Imagine Laurel and Hardy on a teeter-totter — check out Figure 9-5.

Figure 9-5:
Laurel and
Hardy —
blithely
unaware of
the calculus
implications.

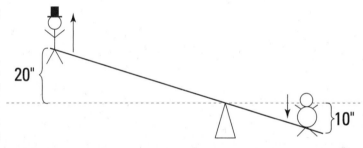

20"

10"

Assuming Hardy weighs twice as much as Laurel, Hardy has to sit twice as close to the center as Laurel for them to balance. And for every inch that Hardy goes down, Laurel goes up two inches. So Laurel moves twice as much as Hardy. Voilà, you've got a derivative!

A derivative is a rate. A *derivative* is simply a measure of how much one thing changes compared to another — and that's a *rate*.

Laurel moves twice as much as Hardy, so with calculus symbols you write

$$dL = 2dH$$

Loosely speaking, *dL* can be thought of as the change in Laurel's position and *dH* as the change in Hardy's position. You can see that if Hardy goes down 10 inches then *dH* is 10, and because *dL* equals 2 times *dH*, *dL* is 20 — so Laurel goes up 20 inches. Dividing both sides of this equation by *dH* gives you

$$\frac{dL}{dH} = 2$$

And that's the derivative of Laurel with respect to Hardy. (It's read as, "dee *L*, dee *H*," or as, "the derivative of *L* with respect to *H*.") The fact that $\frac{dL}{dH} = 2$ simply means that Laurel is moving 2 times as much as Hardy. Laurel's *rate* of movement is 2 inches per inch of Hardy's movement.

Now let's look at it from Hardy's point of view. Hardy moves half as much as Laurel, so you can also write

$$dH = \frac{1}{2} dL$$

Dividing by *dL* gives you

$$\frac{dH}{dL} = \frac{1}{2}$$

This is the derivative of Hardy with respect to Laurel, and it means that Hardy moves $\frac{1}{2}$ inch for every inch that Laurel moves. Thus, Hardy's rate is $\frac{1}{2}$ inch per inch of Laurel's movement. By the way, you can also get this derivative by taking $\frac{dL}{dH} = 2$, which is the same as $\frac{dL}{dH} = \frac{2}{1}$, and flipping it upside down to get $\frac{dH}{dL} = \frac{1}{2}$.

These rates of 2 *inches per inch* and $\frac{1}{2}$ *inch per inch* may seem a bit odd because we often think of rates as referring to something per unit of time, like *miles per hour*. But a rate can be *anything per anything*. So, whenever you've got a *this per that*, you've got a rate; and if you've got a rate, you've got a derivative.

Speed — the most familiar rate

Speaking of *miles per hour,* say you're driving at a constant speed of 60 miles per hour. That's your car's *rate,* and 60 miles per hour is the derivative of your car's position, *p,* with respect to time, *t.* With calculus symbols, you write

$$\frac{dp}{dt} = 60 \frac{\text{miles}}{\text{hour}}$$

This tells you that your car's position changes 60 miles for each hour that the time changes. Or you can say that your car's position (in miles) changes 60 times as much as the time changes (in hours). Again, a derivative just tells you how much one thing changes compared to another.

And just like the Laurel and Hardy example, this derivative, like all derivatives, can be flipped upside down:

$$\frac{dt}{dp} = \frac{1}{60} \frac{\text{hours}}{\text{mile}}$$

This *hours-per-mile* rate is certainly much less familiar than the ordinary *miles-per-hour* rate, but it's nevertheless a perfectly legitimate rate. It tells you that for each mile you go the time changes $\frac{1}{60}$ of an hour. And it tells you that the time (in hours) changes $\frac{1}{60}$ as much as the car's position (in miles).

There's no end to the different rates you might see. We just saw *miles per hour* and *hours per mile*. Then there's *miles per gallon* (for gas mileage), *gallons per minute* (for water draining out of a pool), *output per employee* (for a factory's productivity), and so on. Rates can be constant or changing. In either case, every rate is a derivative, and every derivative is a rate.

The rate-slope connection

Rates and slopes have a simple connection. All of the previous rate examples can be graphed on an *x-y* coordinate system, where each rate appears as a slope. Consider the Laurel and Hardy example again. Laurel moves twice as much as Hardy. This can be represented by the following equation:

$$L = 2H$$

Figure 9-6 shows the graph of this function.

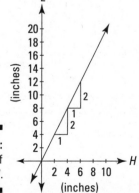

Figure 9-6:
The graph of
$L = 2H$.

The inches on the *H-axis* indicate how far Hardy has moved up or down from the teeter-totter's starting position; the inches on the *L-axis* show how far Laurel has moved up or down. The line goes up 2 inches for each inch it goes to the right, and its slope is thus $\frac{2}{1}$, or 2. This is the visual depiction of $\frac{dL}{dH} = 2$, showing that Laurel's position changes 2 times as much as Hardy's.

One last comment. You know that $slope = \frac{rise}{run}$. Well, you can think of dL as the *rise* and dH as the *run*. That ties everything together quite nicely.

$$slope = \frac{rise}{run} = \frac{dL}{dH} = rate$$

Remember, a derivative is just a slope, and a derivative is also just a rate.

The Derivative of a Curve

The sections so far in this chapter have involved *linear* functions — straight lines with *unchanging* slopes. But if all functions and graphs were lines with unchanging slopes, there'd be no need for calculus. The derivative of the Laurel and Hardy function graphed previously is 2, but you don't need calculus to determine the slope of a line. Calculus is the mathematics of change, so now is a good time to move on to *parabolas,* curves with *changing* slopes. Figure 9-7 is the graph of the parabola, $y = \frac{1}{4}x^2$.

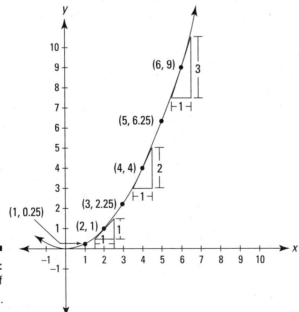

Figure 9-7:
The graph of
$y = \frac{1}{4}x^2$.

Notice how the parabola gets steeper and steeper as you go to the right. You can see from the graph that at the point $(2, 1)$, the slope is 1; at $(4, 4)$, the slope is 2; at $(6, 9)$, the slope is 3, and so on. Unlike the unchanging slope of a line, the slope of a parabola depends on where you are; it depends on the x-coordinate of wherever you are on the parabola. So, the derivative (or slope) of the function $y = \frac{1}{4}x^2$ is itself a function of x — namely $\frac{1}{2}x$ (I show you how I got that in a minute). To find the slope of the curve at any point, you just plug the x-coordinate of the point into the derivative, $\frac{1}{2}x$, and you've got the slope. For instance, if you want the slope at the point $(3, 2.25)$, plug 3 into the x, and the slope is $\frac{1}{2}$ times 3, or 1.5. Table 9-2 shows some points on the parabola and the steepness at those points.

Table 9-2	Points on the Parabola $y = \frac{1}{4}x^2$ and the Slopes at Those Points						
x (horizontal position)	1	2	3	4	5	6	etc.
y (height)	0.25	1	2.25	4	6.25	9	etc.
$\frac{1}{2}x$ (slope)	0.5	1	1.5	2	2.5	3	etc.

Here's the calculus. You write

$$\frac{dy}{dx} = \frac{1}{2}x \text{ or } y' = \frac{1}{2}x$$

And you say,

The derivative of the function $y = \frac{1}{4}x^2$ is $\frac{1}{2}x$.

Or you can say,

The derivative of $\frac{1}{4}x^2$ is $\frac{1}{2}x$.

I promised to tell you how to *derive* this derivative of $y = \frac{1}{4}x^2$, so here you go:

1. **Beginning with the original function, $\frac{1}{4}x^2$, take the power and put it in front of the coefficient.**

   $$2 \cdot \frac{1}{4}x^{②}$$

2. Multiply.

2 times $\frac{1}{4}$ is $\frac{1}{2}$ so that gives you $\frac{1}{2}x^2$.

3. Reduce the power by 1.

In this example, the 2 becomes a 1. So the derivative is

$\frac{1}{2}x^1$ or just $\frac{1}{2}x$.

This and many other differentiation techniques are discussed in Chapter 10.

The Difference Quotient

Sound the trumpets! You come now to what is perhaps the cornerstone of differential calculus: the difference quotient, the bridge between limits and the derivative. (But you're going to have to be patient here, because it's going to take me a few pages to explain the logic behind the difference quotient before I can show you what it is.) Okay, so here goes. I keep repeating — have you noticed? — the important fact that a derivative is just a slope. You learned how to find the slope of a line in algebra. In Figure 9-7, I gave you the slope of the parabola at several points, and then I showed you the short-cut method for finding the derivative — but I left out the important math in the middle. That math involves limits, and it takes us to the threshold of calculus. Hold on to your hat.

Slope **is defined as** $\frac{rise}{run}$, **and** *Slope* $= \frac{y_2 - y_1}{x_2 - x_1}$.

To compute a slope, you need two points to plug into this formula. For a line, this is easy. You just pick any two points on the line and plug them in. But it's not so simple if you want, say, the slope of the parabola $f(x) = x^2$ at the point $(2, 4)$. Check out Figure 9-8.

You can see the line drawn tangent to the curve at $(2, 4)$. Because the slope of the tangent line is the same as the slope of the parabola at $(2, 4)$, all you need is the slope of the tangent line to give you the slope of the parabola. But you don't know the equation of the tangent line, so you can't get the second point — in addition to $(2, 4)$ — that you need for the slope formula.

Here's how the inventors of calculus got around this roadblock. Figure 9-9 shows the tangent line again and a secant line intersecting the parabola at $(2, 4)$ and at $(10, 100)$.

Definition of *secant line*: A secant line is a line that intersects a curve at two points. This is a bit oversimplified, but it'll do.

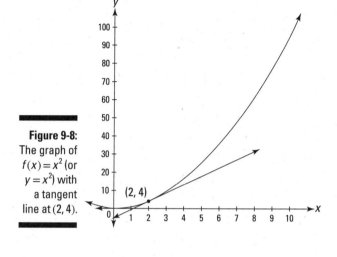

Figure 9-8:
The graph of
$f(x) = x^2$ (or
$y = x^2$) with
a tangent
line at $(2, 4)$.

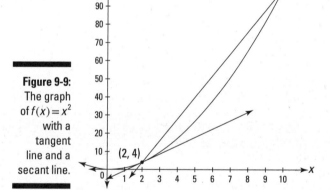

Figure 9-9:
The graph
of $f(x) = x^2$
with a
tangent
line and a
secant line.

The slope of this secant line is given by the slope formula:

$$Slope = \frac{rise}{run} = \frac{y_2 - y_1}{x_2 - x_1} = \frac{100 - 4}{10 - 2} = \frac{96}{8} = 12$$

You can see that this secant line is steeper than the tangent line, and thus the slope of the secant, 12, is higher than the slope you're looking for.

Now add one more point at $(6, 36)$ and draw another secant using that point and $(2, 4)$ again. See Figure 9-10.

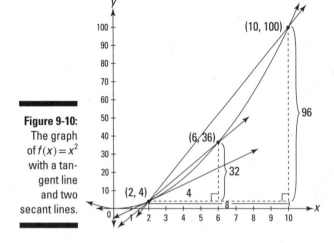

Calculate the slope of this second secant:

$$Slope = \frac{36-4}{6-2} = \frac{32}{4} = 8$$

You can see that this secant line is a better approximation of the tangent line than the first secant.

Now, imagine what would happen if you grabbed the point at $(6, 36)$ and slid it down the parabola toward $(2, 4)$, dragging the secant line along with it. Can you see that as the point gets closer and closer to $(2, 4)$, the secant line gets closer and closer to the tangent line, and that the slope of this secant thus gets closer and closer to the slope of the tangent?

So, you can get the slope of the tangent if you take the *limit* of the slopes of this moving secant. Let's give the moving point the coordinates (x_2, y_2). As this point (x_2, y_2) slides closer and closer to (x_1, y_1), namely $(2, 4)$, the *run*, which equals $x_2 - x_1$, gets closer and closer to zero. So here's the limit you need:

$$Slope_{of \ tangent} = \lim_{\substack{as \ point \ slides \\ toward \ (2,4)}} (slope_{of \ moving \ secant})$$

$$= \lim_{run \to 0} \frac{rise}{run}$$

$$= \lim_{x_2 \to x_1} \frac{y_2 - y_1}{x_2 - x_1}$$

$$= \lim_{x_2 \to 2} \frac{y_2 - 4}{x_2 - 2}$$

Watch what happens to this limit when you plug in four more points on the parabola that are closer and closer to $(2, 4)$:

When the point (x_2, y_2) slides to $(3, 9)$, the slope is $\frac{9-4}{3-2}$, or 5.

When the point slides to $(2.1, 4.41)$, the slope is $\frac{4.41-4}{2.1-2}$, or 4.1.

When the point slides to $(2.01, 4.0401)$, the slope is 4.01.

When the point slides to $(2.001, 4.004001)$, the slope is 4.001.

Sure looks like the slope is headed toward 4. (By the way, the fact that the slope at $(2, 4)$ — which you'll see in a minute does turn out to be 4 — is the same as the y-coordinate of the point is a meaningless coincidence, as is the pattern you may have noticed in the above numbers between the y-coordinates and the slopes.)

As with all limit problems, the variable in this problem, x_2, *approaches* but never actually gets to the arrow-number (2 in this case). If it got to 2 — which would happen if you slid the point you grabbed along the parabola until it was actually on top of $(2, 4)$ — you'd get $\frac{4-4}{2-2} = \frac{0}{0}$, which is undefined. But, of course, the slope at $(2, 4)$ is precisely the slope you want — the slope of the line when the point *does* land on top of $(2, 4)$. Herein lies the beauty of the limit process. With this limit, you get the *exact* slope of the *tangent* line at $(2, 4)$ even though the limit function, $\frac{y_2-4}{x_2-2}$, generates slopes of *secant* lines.

Here again is the equation for the slope of the tangent line:

$$Slope = \lim_{x_2 \to 2} \frac{y_2-4}{x_2-2}$$

And the slope of the tangent line is — you guessed it — the derivative.

Meaning of the *derivative:* The derivative of a function $f(x)$ at some number $x = c$, written as $f'(c)$, is the slope of the tangent line to f drawn at c.

The slope fraction $\frac{y_2-4}{x_2-2}$ is expressed with algebra terminology. Now let's rewrite it to give it that highfalutin calculus look. But first, finally, the definition you've been waiting for.

Definition of the *difference quotient:* There's a fancy calculus term for the general slope fraction, $\frac{rise}{run}$ or $\frac{y_2-y_1}{x_2-x_1}$, when you write it in the fancy calculus way. A fraction is a *quotient,* right? And both y_2-y_1 and x_2-x_1 are *differences,* right? So, voilà, it's called the *difference quotient.* Here it is:

$$\frac{f(x+h)-f(x)}{h}$$

(This is the most common way of writing the difference quotient. You may run across other, equivalent ways.) In the next two pages, I show you how $\frac{y_2 - y_1}{x_2 - x_1}$ morphs into the difference quotient.

Okay, let's lay out this morphing process. First, the *run*, $x_2 - x_1$ (in this example, $x_2 - 2$), is called — don't ask me why — h. Next, because $x_1 = 2$ and the *run* equals h, x_2 equals $2 + h$. You then write y_1 as $f(2)$ and y_2 as $f(2 + h)$. Making all the substitutions gives you the derivative of x^2 at $x = 2$:

$$f'(2) = \lim_{run \to 0} \frac{rise}{run}$$
$$= \lim_{x_2 \to 2} \frac{y_2 - 4}{x_2 - 2}$$
$$= \lim_{h \to 0} \frac{f(2+h) - f(2)}{(2+h) - 2}$$
$$= \lim_{h \to 0} \frac{f(2+h) - f(2)}{h}$$

REMEMBER

$\lim\limits_{h \to 0} \dfrac{f(2+h) - f(2)}{h}$ **is simply the shrinking** $\dfrac{rise}{run}$ **stair step you can see in Figure 9-10 as the point slides down the parabola toward (2, 4).**

Figure 9-11 is basically the same as Figure 9-10 except that instead of exact points like $(6, 36)$ and $(10, 100)$, the sliding point has the general coordinates of $(2+h, f(2+h))$, and the *rise* and the *run* are expressed in terms of h. Figure 9-11 is the ultimate figure for $f'(2) = \lim\limits_{h \to 0} \dfrac{f(2+h) - f(2)}{h}$.

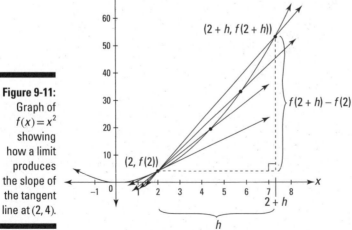

Figure 9-11:
Graph of
$f(x) = x^2$
showing
how a limit
produces
the slope of
the tangent
line at $(2, 4)$.

Have I confused you with these two figures? Don't sweat it. They both show the same thing. Both figures are visual representations of $f'(2) = \lim\limits_{h \to 0} \dfrac{f(2+h) - f(2)}{h}$. I just thought it'd be a good idea to show you a figure with exact coordinates before showing you Figure 9-11 with all that strange-looking f and h stuff in it.

Doing the math gives you, at last, the slope of the tangent line at $(2, 4)$:

$$
\begin{aligned}
f'(2) &= \lim_{h \to 0} \frac{f(2+h) - f(2)}{h} \\[6pt]
&= \lim_{h \to 0} \frac{(2+h)^2 - (2)^2}{h} \quad \text{(The function is } f(x) = x^2, \\
&\phantom{= \lim_{h \to 0}} \qquad\qquad\qquad\quad \text{so } f(2+h) = (2+h)^2, \text{ right?)} \\[6pt]
&= \lim_{h \to 0} \frac{(4 + 4h + h^2) - 4}{h} \\[6pt]
&= \lim_{h \to 0} \frac{4h + h^2}{h} \\[6pt]
&= \lim_{h \to 0} \frac{h(4 + h)}{h} \\[6pt]
&= \lim_{h \to 0} (4 + h) \\[6pt]
&= 4 + 0 = 4
\end{aligned}
$$

So the slope at the point $(2, 4)$ is 4.

Main definition of the *derivative*: If you replace the point $(2, f(2))$ in the limit equation above with the general point $(x, f(x))$, you get the general definition of the derivative as a function of x:

$$
f'(x) = \lim_{h \to 0} \frac{f(x+h) - f(x)}{h}
$$

So at last you see that the derivative is defined as the limit of the difference quotient.

Figure 9-12 shows this general definition graphically. Note that Figure 9-12 is virtually identical to Figure 9-11 except that xs replace the 2s in Figure 9-11 and that the moving point in Figure 9-12 slides down toward any old point $(x, f(x))$ instead of toward the specific point $(2, f(2))$.

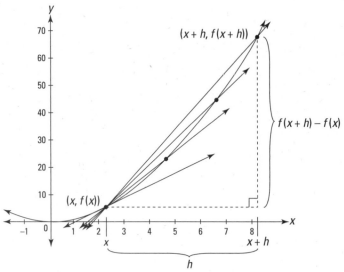

Now work out this limit and get the derivative for the parabola $f(x)=x^2$:

$$f'(x)=\lim_{h\to 0}\frac{f(x+h)-f(x)}{h}$$

$$=\lim_{h\to 0}\frac{(x+h)^2-x^2}{h} \quad \text{(The function is } f(x)=x^2,$$
$$\text{so } f(x+h)=(x+h)^2.)$$

$$=\lim_{h\to 0}\frac{(x^2+2xh+h^2)-x^2}{h}$$

$$=\lim_{h\to 0}\frac{2xh+h^2}{h}$$

$$=\lim_{h\to 0}\frac{h(2x+h)}{h}$$

$$=\lim_{h\to 0}(2x+h)$$

$$=2x+0=2x$$

Thus for this parabola, the derivative (which is the slope of the tangent line at each value x) equals $2x$. Plug any number into x, and you get the slope of the parabola at that x-value. Try it.

To close this section, let's look at one final figure. Figure 9-13 sort of summarizes (in a simplified way) all the difficult preceding ideas about the difference quotient. Like Figures 9-10, 9-11, and 9-12, Figure 9-13 contains a basic slope stair-step, a secant line, and a tangent line. The slope of the secant line is $\frac{rise}{run}$, or $\frac{\Delta y}{\Delta x}$. The slope of the tangent line is $\frac{dy}{dx}$. You can think of $\frac{dy}{dx}$ as $\frac{a\,little\,(ultimately\,infinitesimal)\,bit\,of\,y}{a\,little\,(ultimately\,infinitesimal)\,bit\,of\,x}$, and you can see why this is one of the symbols used for the derivative. As the secant line stair-step shrinks down to nothing, or, in other words, in the limit as Δx and Δy go to zero,

$$\frac{dy}{dx} \text{ (the slope of the tangent line)} = \frac{\Delta y}{\Delta x} \text{ (the slope of the secant line)}.$$

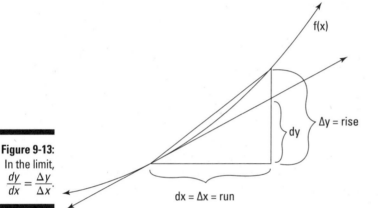

Figure 9-13:
In the limit,
$\frac{dy}{dx} = \frac{\Delta y}{\Delta x}$.

Average Rate and Instantaneous Rate

Returning once again to the connection between slopes and rates, a slope is just the visual depiction of a rate: The slope, $\frac{rise}{run}$, just tells you the rate at which y changes compared to x. If, for example, the y-coordinate tells you distance traveled (in miles), and the x-coordinate tells you elapsed time (in hours), you get the familiar rate of *miles per hour*.

Each secant line in Figures 9-9 and 9-10 has a slope given by the formula $\frac{y_2 - y_1}{x_2 - x_1}$. That slope is the *average* rate over the interval from x_1 to x_2. If y is in miles and x is in hours, you get the *average* speed in *miles per hour* during the time interval from x_1 to x_2.

When you take the limit and get the slope of the tangent line, you get the *instantaneous* rate at the point (x_1, y_1). Again, if y is in miles and x is in hours, you get the *instantaneous* speed at the single point in time, x_1. Because the slope of the tangent line is the derivative, this gives us another definition of the derivative.

Another definition of the *derivative:* The derivative of a function $f(x)$ at some x-value is the *instantaneous* rate of change of f with respect to x at that value.

To Be or Not to Be? Three Cases Where the Derivative Does Not Exist

I want to discuss the three situations where a derivative fails to exist (see the "33333 Limit Mnemonic" section in Chapter 7). By now you certainly know that the derivative of a function at a given point is the slope of the tangent line at that point. So, if you can't draw a tangent line, there's no derivative — that happens in the first two cases below. In the third case, there's a tangent line, but its slope and the derivative are undefined.

✔ There's no tangent line and thus no derivative at any type of *discontinuity:* removable, infinite, or jump. (These types of discontinuity are discussed and illustrated in Chapter 7.) Continuity is, therefore, a *necessary* condition for differentiability. It's not, however, a *sufficient* condition as the next two cases show. Dig that logician-speak.

✔ There's no tangent line and thus no derivative at a sharp *corner* on a function (or at a *cusp*, a really pointy, sharp turn). See function f in Figure 9-14.

✔ Where a function has a *vertical tangent line* (which occurs at a vertical inflection point), the slope is undefined, and thus the derivative fails to exist. See function g in Figure 9-14. (Inflection points are explained in Chapter 11.)

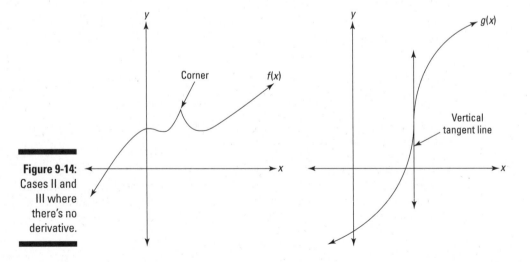

Figure 9-14: Cases II and III where there's no derivative.

Chapter 10

Differentiation Rules — Yeah, Man, It Rules

. . . .

. . . .

Chapter 9 gives you the basic idea of what a derivative is — it's just a rate like speed and it's simply the slope of a function. It's important that you have a solid, intuitive grasp of these fundamental ideas.

You also now know the mathematical foundation of the derivative and its technical definition involving the limit of the difference quotient. Now, I'm going to be forever banned from the Royal Order of Pythagoras for saying this, but, to be perfectly candid, you can basically forget that limit stuff — except that you need to know it for your final — because in this chapter I give you shortcut techniques for finding derivatives that avoid the difficulties of limits and the difference quotient.

Some of this material is unavoidably dry. If you have trouble staying awake while slogging through these rules, look back to the last chapter and take a peek at the next three chapters to see why you should care about mastering these differentiation rules. Countless problems in business, economics, medicine, engineering, and physics, as well as other disciplines, deal with how fast a function is rising or falling, and that's what a derivative tells us. And it's often important to know where a function is rising or falling the fastest (the largest and smallest slopes) and where its peaks and valleys are (where the slope is zero). Before you can do these interesting problems, you've got to learn how to find derivatives. If Chapters 11, 12, and 13 are like playing the piano, then this chapter is like learning your scales — it's dull, but you've got to do it. You may want to order up a latte with an extra shot.

Basic Differentiation Rules

Calculus can be difficult, but you'd never know it judging by this section alone. Learning these first half dozen or so rules is a snap. If you get tired of this easy stuff, however, I promise plenty of challenges in the next section.

The constant rule

This is simple. $f(x)=5$ is a horizontal line with a slope of zero, and thus its derivative is also zero. So, for any number c, if $f(x)=c$, then $f'(x)=0$. Or you can write $\frac{d}{dx}c=0$. End of story.

The power rule

Say $f(x)=x^5$. To find its derivative, take the power, 5, bring it in front of the x, and then reduce the power by 1 (in this example, the power becomes a 4). That gives you $f'(x)=5x^4$. To repeat, bring the power in front, then reduce the power by 1. That's all there is to it.

In Chapter 9, I differentiated $y=x^2$ with the difference quotient:

$$y=x^2$$
$$y'=\lim_{h\to 0}\frac{(x+h)^2-x^2}{h}$$
$$=\lim_{h\to 0}\frac{x^2+2xh+h^2-x^2}{h}$$
$$=\lim_{h\to 0}\frac{2xh+h^2}{h}$$
$$=\lim_{h\to 0}(2x+h)$$
$$=2x$$

That takes some doing. Instead of all that, just use the power rule: Bring the 2 in front, reduce the power by 1, which leaves you with a power of 1 that you can drop (because a power of 1 does nothing). Thus,

$$y=x^2$$
$$y'=2x$$

Because this is so simple, you may be wondering why we didn't skip the complicated difference quotient stuff and just go straight to the shortcut method. Well, admittedly, that would have saved some time, especially considering the fact that once you know this and other shortcut methods, you'll never need the difference quotient again — except for your final exam. But the difference quotient is included in every calculus book and course because it gives you a fuller, richer understanding of calculus and its foundations — think of it as a mathematical character builder. *Or* because math teachers are sadists. You be the judge.

The power rule works for any power: a positive, a negative, or a fraction:

If $f(x)=x^{-2}$ then $f'(x)=-2x^{-3}$
If $g(x)=x^{2/3}$ then $g'(x)=\frac{2}{3}x^{-1/3}$
If $h(x)=x$ then $h'(x)=1$

The derivative of x is 1. Make sure you remember how to do the derivative of the last function in the above list. It's the simplest of these functions, yet the easiest one to miss.

The best way to understand this derivative is to realize that $h(x)=x$ is a line that fits the form $y=mx+b$ because $h(x)=x$ is the same as $h(x)=1x+0$ (or $y=1x+0$). The slope (m) of this line is 1, so the derivative equals 1. Or you can just memorize that the derivative of x is 1. But if you forget both of these ideas, you can always use the power rule. Rewrite $h(x)=x$ as $h(x)=x^1$, then apply the rule: Bring the 1 in front and reduce the power by 1 to zero, giving you $h'(x)=1x^0$. Because x^0 equals 1, you've got $h'(x)=1$.

Rewrite functions so you can use the power rule. You can differentiate radical functions by rewriting them as power functions and then using the power rule. For example, if $f(x)=\sqrt[3]{x^2}$, rewrite it as $f(x)=x^{2/3}$ and use the power rule. You can also use the power rule to differentiate functions like $f(x)=\frac{1}{x^3}$. Rewrite this as $f(x)=x^{-3}$, then use the power rule.

The constant multiple rule

What if the function you're differentiating begins with a coefficient? Makes no difference. A coefficient has no effect on the process of differentiation. You just ignore it and differentiate according to the appropriate rule. The coefficient stays where it is until the final step when you simplify your answer by multiplying by the coefficient.

Differentiate $y = 4x^3$.

Solution: You know by the power rule that the derivative of x^3 is $3x^2$, so the derivative of $4(x^3)$ is $4(3x^2)$. The 4 just sits there doing nothing. Then, as a final step, you simplify: $4(3x^2)$ equals $12x^2$. So $y' = 12x^2$. (By the way, most people just bring the 3 to the front, like this: $y' = 3 \cdot 4x^2$, which gives you the same result.)

Differentiate $y = 5x$.

Solution: This is a line of the form $y = mx + b$ with $m = 5$, so the slope is 5, and thus the derivative is 5: $y' = 5$. (It's important to think graphically like this from time to time.) But you can also solve the problem with the power rule: $\frac{d}{dx} x^1 = 1x^0 = 1$; so $\frac{d}{dx} 5(x^1) = 5(1) = 5$.

One final example: Differentiate $y = \frac{5x^{1/3}}{4}$.

Solution: The coefficient here is $\frac{5}{4}$. So, because $\frac{d}{dx} x^{1/3} = \frac{1}{3} x^{-2/3}$ (by the power rule), $\frac{d}{dx} \frac{5}{4}(x^{1/3}) = \frac{5}{4}\left(\frac{1}{3} x^{-2/3}\right) = \frac{5}{12} x^{-2/3}$.

pi, e, c, k, etc. are *not* variables! Don't forget that π (~3.14) and e (~2.72) are numbers, not variables, so they behave like ordinary numbers. Constants in problems, like c and k also behave like ordinary numbers. (By the way, the number e, named for the great mathematician Leonhard Euler, is perhaps the most important number in all of mathematics, but I don't get into that here.)

Thus, if $y = \pi x$, $y' = \pi$ — this works exactly like differentiating $y = 5x$. And because π^3 is just a number, if $y = \pi^3$ then $y' = 0$ — this works exactly like differentiating $y = 10$. You'll also see problems containing constants like c and k. Be sure to treat them like regular numbers. For example, the derivative of $y = 5x + 2k^3$ (where k is a constant) is 5, not $5 + 6k^2$.

The sum rule — hey, that's some rule you got there

When you want the derivative of a sum of terms, take the derivative of each term separately.

What's $f'(x)$ if $f(x) = x^6 + x^3 + x^2 + x + 10$?

Solution: Just use the power rule for each of the first four terms and the constant rule for the final term. Thus, $f'(x) = 6x^5 + 3x^2 + 2x + 1$.

The difference rule — it makes no difference

If you've got a difference (that's subtraction) instead of a sum, it makes no difference. You still differentiate each term separately. Thus, if $y = 3x^5 - x^4 - 2x^3 + 6x^2 + 5x$, then $y' = 15x^4 - 4x^3 - 6x^2 + 12x + 5$. The addition and subtraction signs are unaffected by the differentiation.

Differentiating trig functions

Ladies and gentlemen: I have the high honor and distinct privilege of introducing you to the derivatives of the six trig functions:

$$\frac{d}{dx}\sin x = \cos x \qquad \frac{d}{dx}\cos x = -\sin x$$

$$\frac{d}{dx}\tan x = \sec^2 x \qquad \frac{d}{dx}\cot x = -\csc^2 x$$

$$\frac{d}{dx}\sec x = \sec x \tan x \qquad \frac{d}{dx}\csc x = -\csc x \cot x$$

Make sure you memorize the derivatives of sine and cosine. They're a snap, and I've never known anyone to forget them. If you're good at rote memorization, memorize the other four as well. Or, if you're not wild about memorization or are afraid that this knowledge will crowd out the date of the Battle of Hastings (1066) — which is much more likely to come up in a board game than trig derivatives — you can figure out the last four derivatives from scratch by using the quotient rule (see the section "The quotient rule" later on). A third option is to use the following mnemonic trick.

Psst, what's the derivative of cosecant? Imagine you're taking a test and can't remember those four last trig derivatives. You lean over to the guy next to you and whisper, "Psst, what's the derivative of csc *x*?" Now, the last three letters of *psst* (sst) are the initial letters of sec, sec, tan. Write these three down, and below them write their cofunctions: csc, csc, cot. Put a negative sign on the csc in the middle. Finally, add arrows like in the following diagram:

$$\begin{array}{ccccc} \text{sec} & \rightarrow & \text{sec} & \leftarrow & \text{tan} \\ \text{csc} & \rightarrow & -\text{csc} & \leftarrow & \text{cot} \end{array}$$

(This may seem complicated, but, take my word for it, you'll remember the word *psst,* and after that the diagram is very easy to remember.)

Look at the top row. The *sec* on the left has an arrow pointing to *sec tan* — so the derivative of sec*x* is sec*x* tan*x*. The *tan* on the right has an arrow pointing to *sec sec,* so the derivative of tan*x* is sec²*x*. The bottom row works the same way except that both derivatives are negative.

Differentiating exponential and logarithmic functions

Caution: Memorization ahead.

Exponential functions

If you can't memorize the next rule, hang up your calculator.

$$\frac{d}{dx}e^x = e^x$$

That's right — break out the smelling salts — the derivative of e^x is itself! This is a special function: e^x and its multiples, like $5e^x$, are the only functions that are their own derivatives. Think about what this means. Look at the graph of $y = e^x$ in Figure 10-1.

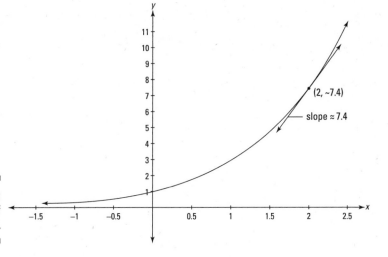

Figure 10-1:
The graph of
$y = e^x$.

Pick any point on this function, say $(2, \sim 7.4)$, and the height of the function at that point, ~ 7.4, is the same as the slope at that point.

If the base is a number other than e, you have to tweak the derivative by multiplying it by the natural log of the base:

If $y = 2^x$ then $y' = 2^x \ln 2$
If $y = 10^x$ then $y' = 10^x \ln 10$

Logarithmic functions

And now — what you've all been waiting for — the derivatives of logarithmic functions. (See Chapter 4 if you want to brush up on logs.) Here's the derivative of the *natural* log — that's the log with base e:

$$\frac{d}{dx} \ln x = \frac{1}{x}$$

If the log base is a number other than e, you tweak this derivative — like with exponential functions — except that you *divide* by the natural log of the base instead of multiplying. Thus,

$$\frac{d}{dx} \log_2 x = \frac{\frac{1}{x}}{\ln 2} = \frac{1}{x \ln 2}, \text{ and}$$

$$\frac{d}{dx} \log x = \frac{1}{x \ln 10} \quad \text{(Recall that } \log x \text{ means } \log_{10} x, \text{ so the base is 10.)}$$

Differentiation Rules for Experts — Oh, Yeah, I'm a Calculus Wonk

Now that you've *totally* mastered all the basic rules, take a breather and rest on your laurels for a minute. . . . Okay, ready for a challenge? The following rules, especially the chain rule, can be tough. But you know what they say: "No pain, no gain," "No guts, no glory," yada, yada, yada.

The product rule

You use this rule for — hold on to your hat — the *product* of two functions like

$$y = x^3 \cdot \sin x$$

The product rule:

If $y = this \cdot that$,
then $y' = this' \cdot that + this \cdot that'$

So, for $y = x^3 \cdot \sin x$,

$$y' = (x^3)' \cdot \sin x + x^3 \cdot (\sin x)'$$
$$= 3x^2 \sin x + x^3 \cos x$$

The quotient rule

I have a feeling that you can guess what this rule is for — the *quotient* of two functions like

$$y = \frac{\sin x}{x^4}$$

The quotient rule:

If $y = \dfrac{top}{bottom}$,

then $y' = \dfrac{top' \cdot bottom - top \cdot bottom'}{bottom^2}$

Just about every calculus book I've ever seen gives this rule in a slightly different form that's harder to remember. And some books give a "mnemonic" involving the words *lodeehi* and *hideelo* or *hodeehi* and *hideeho,* which is very easy to get mixed up — great, thanks a lot.

Memorize the quotient rule the way I've written it. You'll have no problem remembering what goes in the denominator — no one ever forgets it. The trick is knowing the order of the terms in the numerator. Think of it like this: You're doing a derivative, so the first thing you do is to take a derivative. And is it more natural to begin at the top or the bottom of a fraction? The top, of course. So the quotient rule begins with the derivative of the top. If you remember that, the rest of the numerator is almost automatic. (Note that the product rule begins with the derivative of the first function you read as you read the product of two functions from left to right. In the same way, the quotient rule begins with the derivative of the first function you read as you read the quotient of two functions from top to bottom.) Focus on these points and you'll remember the quotient rule ten years from now — oh, sure.

So here's the derivative of $y = \dfrac{\sin x}{x^4}$:

$$y' = \frac{(\sin x)' \cdot x^4 - \sin x \cdot (x^4)'}{(x^4)^2}$$

$$= \frac{x^4 \cos x - 4x^3 \sin x}{x^8}$$

$$= \frac{x^3(x \cos x - 4 \sin x)}{x^8}$$

$$= \frac{x \cos x - 4 \sin x}{x^5}$$

In the "Differentiating trig functions" section, I promised to show you how to find the derivatives of four trig functions — *tangent, cotangent, secant,* and *cosecant* — with the quotient rule. I'm a man of my word, so here goes. All four of these functions can be written in terms of *sine* and *cosine*, right? (See Chapter 6.) For instance, $\tan x = \dfrac{\sin x}{\cos x}$. Now, if you want the derivative of $\tan x$, you can use the quotient rule:

$$\tan x = \frac{\sin x}{\cos x}$$

$$(\tan x)' = \frac{(\sin x)' \cdot \cos x - \sin x \cdot (\cos x)'}{\cos^2 x}$$

$$= \frac{\cos x \cdot \cos x - \sin x \cdot (-\sin x)}{\cos^2 x}$$

$$= \frac{\cos^2 x + \sin^2 x}{\cos^2 x}$$

$$= \frac{1}{\cos^2 x} \qquad \text{(The Pythagorean identity tells}$$
$$\text{you that } \cos^2 x + \sin^2 x = 1.)$$

$$= \sec^2 x$$

Granted, this is quite a bit of work compared to just memorizing the answer or using the mnemonic device presented several pages back, but it's nice to know that you can get the answer this way as a last resort. The other three functions are no harder. Give them a try.

The chain rule

The chain rule is by far the trickiest derivative rule, but it's not really that bad if you carefully focus on a few important points. Let's begin by differentiating $y = \sqrt{4x^3 - 5}$. You use the chain rule here because you've got a *composite* function, that's one function $(4x^3 - 5)$ inside another function (the square root function).

How to spot a *composite* function: $y = \sqrt{x}$ is *not* a composite function because the *argument* of the square root function — that's the thing you take the square root of — is simply x. Whenever the argument of a function is anything other than a plain old x, you've got a composite function. Be careful to distinguish a composite function from something like $y = \sqrt{x} \cdot \sin x$, which is the *product* of two functions, $\sqrt{x}$ and $\sin x$, each of which *does* have just a plain old x as its argument.

Okay, so you've got this composite function, $y = \sqrt{4x^3 - 5}$. Here's how to differentiate it with the chain rule:

1. **You start with the *outside* function, $\sqrt{}$, and differentiate that, IGNORING what's inside. To make sure you ignore the inside, temporarily replace the inside function with the word *stuff*.**

   So you've got $y = \sqrt{stuff}$. Okay, now differentiate $y = \sqrt{stuff}$ the same way you'd differentiate $y = \sqrt{x}$. Because $y = \sqrt{x}$ is the same as $y = x^{1/2}$, the power rule gives you $y' = \frac{1}{2}x^{-1/2}$. So for this problem, you begin with $\frac{1}{2}stuff^{-1/2}$.

2. **Multiply the result from Step 1 by the derivative of the inside function, *stuff'*.**

   $$y' = \frac{1}{2}stuff^{-1/2} \cdot stuff'$$

   Take a good look at this. *All* chain rule problems follow this basic idea. You do the derivative rule for the outside function, ignoring the inside *stuff*, then multiply that by the derivative of the *stuff*.

3. **Differentiate the inside *stuff*.**

   The inside *stuff* in this problem is $4x^3 - 5$ and its derivative is $12x^2$ by the power rule.

4. **Now put the real *stuff* and its derivative back where they belong.**

$$y' = \frac{1}{2}(4x^3 - 5)^{-1/2} \cdot (12x^2)$$

5. **Simplify.**

$$y' = 6x^2(4x^3 - 5)^{-1/2}$$

Or, if you've got something against negative powers, $y' = \dfrac{6x^2}{(4x^3 - 5)^{1/2}}$.

Or, if you've got something against fraction powers, $y' = \dfrac{6x^2}{\sqrt{4x^3 - 5}}$.

Let's try differentiating another composite function, $y = \sin(x^2)$:

1. **The outside function is the sine function, so you start there, taking the derivative of sine and ignoring the inside *stuff*, x^2. The derivative of sin x is cos x, so for this problem, you begin with**

   $\cos(stuff)$

2. **Multiply the derivative of the outside function by the derivative of the *stuff*.**

$$y' = \cos(stuff) \cdot stuff'$$

3. **The *stuff* in this problem is x^2, so *stuff'* is $2x$. When you plug these terms back in, you get**

$$y' = \cos(x^2) \cdot 2x$$
$$= 2x \cos(x^2)$$

Sometimes figuring out which function is inside which can be a bit tricky — especially when a function is inside another and then both of them are inside a *third* function (you can have four or more nested functions, but three is probably the most you'll see). Here's a tip.

Parentheses are your friend. For chain rule problems, rewrite a composite function with a set of parentheses around each inside function, and rewrite trig functions like $\sin^2 x$ with the power outside a set of parentheses: $(\sin x)^2$.

For example — this is a tough one, gird your loins — differentiate $y = \sin^3(5x^2 - 4x)$. First, rewrite the cubed sine function: $y = (\sin(5x^2 - 4x))^3$. Now it's easy to see the order in which the functions are nested. The innermost function is inside the innermost parentheses — that's $5x^2 - 4x$. Next, the sine function is inside the next set of parentheses — that's sin(*stuff*). Last, the cubing function is on the outside of everything — that's *stuff*^{3}. (Because I'm a math teacher, I'm honor bound to point out that the *stuff* in *stuff*^{3} is different from the *stuff* in sin(*stuff*). It's quite unmathematical of me to use the same term to refer to different things, but don't sweat it. I'm just using the term *stuff* to refer to whatever is inside any function.) Okay, now that you know the order of the functions, you can differentiate from *outside in:*

1. **The outermost function is *stuff*^{3} and its derivative is given by the power rule.**

    $3stuff^2$

2. **As with all chain rule problems, you multiply that by *stuff*′.**

    $3stuff^2 \cdot stuff'$

3. **Now put the *stuff*, $\sin(5x^2 - 4x)$, back where it belongs.**

    $3(\sin(5x^2 - 4x))^2 \cdot (\sin(5x^2 - 4x))'$

4. **Use the chain rule again.**

    You can't finish this problem quickly by just taking a simple derivative because you have to differentiate another composite function, $\sin(5x^2 - 4x)$. Just treat $\sin(5x^2 - 4x)$ as if it were the original problem and take its derivative. The derivative of sin x is cos x, so the derivative of sin(*stuff*) begins with cos(*stuff*). Multiply that by *stuff*′. Thus, the derivative of sin(*stuff*) is

    $\cos(stuff) \cdot stuff'$

5. **The *stuff* for this step is $5x^2 - 4x$ and its derivative is $10x - 4$. Plug those things back in.**

    $\cos(5x^2 - 4x) \cdot (10x - 4)$

6. **Now that you've got the derivative of $\sin(5x^2 - 4x)$, plug this result into the result from Step 3, giving you the whole enchilada.**

$$3(\sin(5x^2 - 4x))^2 \cdot \cos(5x^2 - 4x) \cdot (10x - 4)$$

7. **This can be simplified a bit.**

$$(30x - 12)\sin^2(5x^2 - 4x)\cos(5x^2 - 4x)$$

I told you it was a tough one.

It may have occurred to you that you can save some time by not switching to the word *stuff* and then switching back. That's true, but some people like to use the technique because it forces them to leave the *stuff* alone during each step of a problem. That's the critical point.

Make sure you . . . DON'T TOUCH THE *STUFF*.

As long as you remember this, you don't need to actually use the word *stuff* when doing a chain rule problem. You've just got to be sure you don't change the inside function while differentiating the outside function. Say you want to differentiate $f(x) = \ln(x^3)$. The argument of this natural logarithm function is x^3. Don't touch it during the first step of the solution, which is to use the natural log rule: $\frac{d}{dx}\ln x = \frac{1}{x}$. This rule tells you to put the argument of the function in the denominator under the number 1. So, after the first step in differentiating $\ln(x^3)$, you've got $\frac{1}{x^3}$. You then finish the problem by multiplying that by the derivative of x^3 which is $3x^2$. Final answer after simplifying: $\frac{3}{x}$.

With the chain rule, don't use two derivative rules at the same time.
Another way to make sure you've got the chain rule straight is to remember that you never use more than one derivative rule at a time.

In the example above, $\ln(x^3)$, you first use the natural log rule, then, as a *separate step,* you use the power rule to differentiate x^3. At no point in any chain rule problem do you use both rules at the same time. For example, with $\ln(x^3)$, you do not use the natural log rule and the power rule at the same time to come up with $\frac{1}{3x^2}$.

Here's the chain rule mumbo jumbo.

The chain rule (for differentiating a composite function):

If $y = f(g(x))$,
then $y' = f'(g(x)) \cdot g'(x)$

Or, equivalently,

If $y = f(u)$ and $u = g(x)$,

then $\dfrac{dy}{dx} = \dfrac{dy}{du} \cdot \dfrac{du}{dx}$ (Notice how the *du*s cancel.)

See the sidebar, "Why the chain rule works," for a plain-English explanation of this mumbo jumbo.

Why the chain rule works

You wouldn't know it from the difficult math in this section or the fancy chain rule mumbo jumbo, but the chain rule is based on a *very* simple idea. Say one person is walking, another jogging, and a third is riding a bike. If the biker goes four times as fast as the jogger, and the jogger goes twice as fast as the walker, then the biker goes 4 times 2, or 8 times as fast as the walker, right? That's the chain rule in a nutshell — you just multiply the relative rates.

Remember Figure 9-5 showing Laurel and Hardy on a teeter-totter? Recall that for every inch Hardy goes down, Laurel goes up 2 inches. So, Laurel's rate of movement is twice Hardy's rate, and therefore $\dfrac{dL}{dH} = 2$. Now imagine that Laurel has one of those party favors in his mouth (the kind that unrolls as you blow into it) and that for every inch he goes up, he blows the noisemaker out 3 inches. The rate of movement of the noisemaker (*N*) is thus 3 times Laurel's rate of movement. In calculus symbols, $\dfrac{dN}{dL} = 3$. So, how fast is the noisemaker moving compared to Hardy? This is just common sense. The noisemaker is moving 3 times as fast as Laurel, and Laurel is moving 2 times as fast as Hardy, so the noisemaker is moving 3 times 2, or 6 times as fast as Hardy. Here it is in symbols (note that this is the same as the formal definition of the chain rule next to the Mumbo Jumbo icon):

$$\frac{dN}{dH} = \frac{dN}{dL} \cdot \frac{dL}{dH} = 3 \cdot 2 = 6$$

Mere child's play.

One final example and one last tip. Differentiate $4x^2\sin(x^3)$. This problem has a new twist — it involves the chain rule *and* the product rule. How should you begin?

Where do I begin? If you're not sure where to begin differentiating a complex expression, imagine plugging a number into x and then evaluating the expression on your calculator one step at a time. Your *last* computation tells you the *first* thing to do.

Say you plug the number 5 into the xs in $4x^2\sin(x^3)$. You evaluate $4 \cdot 5^2$ — that's 100; then, after getting $5^3 = 125$, you do $\sin(125)$, which is about -0.616. Finally, you multiply 100 by -0.616. Because your *last* computation is *multiplication,* your *first* step in differentiating is to use the *product* rule. (Had your last computation been instead something like $\sin(125)$, then you'd begin with the chain rule.) Remember the product rule?

The product rule:

If $y = this \cdot that$, then $y' = this' \cdot that + this \cdot that'$.

So for $f(x) = 4x^2\sin(x^3)$,

$$f'(x) = (4x^2)'(\sin(x^3)) + (4x^2)(\sin(x^3))'$$

You finish the problem by taking the derivative of $4x^2$ with the power rule and the derivative of $\sin(x^3)$ with the chain rule:

$$f'(x) = (8x)(\sin(x^3)) + (4x^2)(\cos(x^3) \cdot 3x^2)$$

And now simplify:

$$f'(x) = 8x\sin(x^3) + 12x^4\cos(x^3)$$

Differentiating Implicitly

All the differentiation problems presented in previous sections of this chapter are functions like $y = x^2 + 5x$ or $y = \sin x$. In such cases, y is written *explicitly* as a function of x. This means that the equation is solved for y; in other words, y is by itself on one side of the equation. (Note that y was sometimes written as $f(x)$ as in $f(x) = x^3 - 4x^2$, but remember that that's the same thing as $y = x^3 - 4x^2$.)

Sometimes, however, you are asked to differentiate an equation that's not solved for y, like $y^5 + 3x^2 = \sin x - 4y^3$. This equation defines y *implicitly* as a function of x, and you can't write it as an explicit function because it can't be solved for y. For such a problem, you need *implicit differentiation.* When differentiating implicitly, all the derivative rules work the same, with one exception: When you differentiate a term with a y in it, you use the chain rule with a little twist.

Remember using the chain rule to differentiate something like $\sin(x^3)$ with the *stuff* technique? The derivative of sine is cosine, so the derivative of $\sin(stuff)$ is $\cos(stuff) \cdot stuff'$. You finish the problem by finding the derivative of the *stuff*, x^3, which is $3x^2$, and then making the substitutions to give you $\cos(x^3) \cdot 3x^2$. With implicit differentiation, a y works like the word *stuff*. Thus, because

$$(\sin(stuff))' = \cos(stuff) \cdot stuff',$$
$$(\sin y)' = \cos y \cdot y'.$$

The twist is that while the word *stuff* is temporarily taking the place of some *known* function of x (x^3 in this example), y is some *unknown* function of x (you don't know what the y equals in terms of x). And because you don't know what y equals, the y and the y' — unlike the *stuff* and the *stuff'*— must remain in the final answer. But the concept is exactly the same, and you treat y just like the *stuff*. You just can't make the switch back to xs at the end of the problem like you can with a regular chain rule problem.

I suppose you're wondering whether I'm ever going to get around to actually doing the problem. Here goes. Again, differentiate $y^5 + 3x^2 = \sin x - 4y^3$:

1. **Differentiate each term on *both* sides of the equation.**

   $$y^5 + 3x^2 = \sin x - 4y^3$$

   For the first and fourth terms, you use the power rule and, because these terms contain ys, you also use the chain rule. For the second term, you use the regular power rule. And for the third term, you use the regular sine rule.

   $$5y^4 \cdot y' + 6x = \cos x - 12y^2 \cdot y'$$

2. **Collect all terms containing a y' on the left side of the equation and all other terms on the right side.**

   $$5y^4 \cdot y' + 12y^2 \cdot y' = \cos x - 6x$$

3. **Factor out y'.**

   $$y'(5y^4 + 12y^2) = \cos x - 6x$$

4. Divide for the final answer.

$$y' = \frac{\cos x - 6x}{5y^4 + 12y^2}$$

Note that this derivative, unlike the others you've done so far, is expressed in terms of x and y instead of just x. So, if you want to evaluate the derivative to get the slope at a particular point, you need to have values for both x and y to plug into the derivative.

Also note that in many textbooks, the symbol $\frac{dy}{dx}$ is used instead of y' in every step of solutions like the one above. I find y' easier and less cumbersome to work with. But $\frac{dy}{dx}$ does have the advantage of reminding you that you're finding the derivative of y with respect to x. Either way is fine. Take your pick.

Getting into the Rhythm with Logarithmic Differentiation

Say you want to differentiate $f(x) = (x^3 - 5)(3x^4 + 10)(4x^2 - 1)(2x^5 - 5x^2 - 10)$. Now, you could multiply the whole thing out and then differentiate, but that would be a *huge* pain. Or you could use the product rule a few times, but that would also be too tedious and time-consuming. The better way is to use logarithmic differentiation:

1. **Take the natural log of both sides.**

   $$\ln f(x) = \ln\left((x^3 - 5)(3x^4 + 10)(4x^2 - 1)(2x^5 - 5x^2 - 10)\right)$$

2. **Now use the property for the log of a product, which you remember of course (if not, see Chapter 4).**

   $$\ln f(x) = \ln(x^3 - 5) + \ln(3x^4 + 10) + \ln(4x^2 - 1) + \ln(2x^5 - 5x^2 - 10)$$

3. **Differentiate both sides.**

   According to the chain rule, the derivative of $\ln f(x)$ is $\frac{1}{f(x)} \cdot f'(x)$, or $\frac{f'(x)}{f(x)}$.

   (The $f(x)$ works just like the word *stuff* in a regular chain rule problem or a y in an implicit differentiation problem.) For each of the four terms on the right side of the equation, you use the chain rule:

   $$\frac{f'(x)}{f(x)} = \frac{3x^2}{(x^3 - 5)} + \frac{12x^3}{(3x^4 + 10)} + \frac{8x}{(4x^2 - 1)} + \frac{10x^4 - 10x}{(2x^5 - 5x^2 - 10)}$$

4. Multiply both sides by $f(x)$ and you're done.

$$f'(x) = \left(\frac{3x^2}{(x^3-5)} + \frac{12x^3}{(3x^4+10)} + \frac{8x}{(4x^2-1)} + \frac{10x^4-10x}{(2x^5-5x^2-10)} \right) \cdot$$
$$(x^3-5)(3x^4+10)(4x^2-1)(2x^5-5x^2-10)$$

(***Note:*** Make sure you read this monster equation correctly. The right side of the first line gets multiplied by the second line.)

Granted, this answer is pretty hairy, and the solution process isn't exactly a walk in the park, but, take my word for it, this method is *much* easier than the other alternatives.

Differentiating Inverse Functions

There's a difficult-looking formula involving the derivatives of inverse functions, but before we get to it, look at Figure 10-2, which nicely sums up the whole idea.

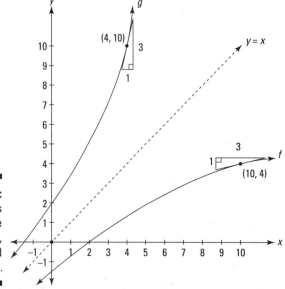

Figure 10-2:
The graphs
of inverse
functions,
$f(x)$ and
$g(x)$.

Figure 10-2 shows a pair of inverse functions, f and g. Recall that inverse functions are symmetrical with respect to the line, $y = x$. As with any pair of inverse functions, if the point $(10, 4)$ is on one function, $(4, 10)$ is on its inverse. And, because of the symmetry of the graphs, you can see that the slopes at those points are reciprocals: At $(10, 4)$ the slope is $\frac{1}{3}$ and at $(4, 10)$ the slope is $\frac{3}{1}$. That's how the idea works graphically, and if you're with me so far, you've got it down at least visually.

The algebraic explanation is a bit trickier, however. The point $(10, 4)$ on f can be written as $(10, f(10))$ and the slope at this point — and thus the derivative — can be expressed as $f'(10)$. The point $(4, 10)$ on g can be written as $(4, g(4))$. Then, because $f(10) = 4$, you can replace the 4s in $(4, g(4))$ with $f(10)$s giving you $(f(10), g(f(10)))$. The slope and derivative at this point can be expressed as $g'(f(10))$. These two slopes are reciprocals, so that gives you the equation

$$f'(10) = \frac{1}{g'(f(10))}$$

This difficult equation expresses nothing more and nothing less than the two triangles on the two functions in Figure 10-2.

Using x instead of 10 gives you the general formula:

The derivative of an inverse function: If f and g are inverse functions, then

$$f'(x) = \frac{1}{g'(f(x))}$$

In words, this formula says that the derivative of a function, f, with respect to x, is the reciprocal of the derivative of its inverse function with respect to f.

Okay, so maybe it was *a lot* trickier.

Scaling the Heights of Higher Order Derivatives

Finding a second, third, fourth, or higher derivative is incredibly simple. The second derivative of a function is just the derivative of its first derivative. The third derivative is the derivative of the second derivative, the fourth derivative is the derivative of the third, and so on. For example, here's a function and its first, second, third, and subsequent derivatives. In this example, all the derivatives are obtained by the power rule:

$$f(x) = x^4 - 5x^2 + 12x - 3$$
$$f'(x) = 4x^3 - 10x + 12$$
$$f''(x) = 12x^2 - 10$$
$$f'''(x) = 24x$$
$$f^{(4)}(x) = 24$$
$$f^{(5)}(x) = 0$$
$$f^{(6)}(x) = 0$$
$$etc. = 0$$
$$etc. = 0$$

All polynomial functions like this one eventually go to zero when you differentiate repeatedly. Rational functions like $f(x) = \dfrac{x^2 - 5}{x + 8}$, on the other hand, get messier and messier as you take higher and higher derivatives. And the higher derivatives of sine and cosine are cyclical. For example,

$$y = \sin x$$
$$y' = \cos x$$
$$y'' = -\sin x$$
$$y''' = -\cos x$$
$$y^{(4)} = \sin x$$

The cycle repeats indefinitely with every multiple of four.

In Chapters 11 and 12, I show you several uses of higher derivatives — mainly second derivatives. (Here's a sneak preview: The first derivative of position is velocity, and the second derivative of position is acceleration.) But for now, let me give you just one of the main ideas in a nutshell. A first derivative, as you know, tells you how fast a function is changing — how fast it's going up or down — that's its slope. A second derivative tells you how fast the first derivative is changing — or, in other words, how fast the slope is changing. A third derivative tells you how fast the second derivative is changing, which tells you how fast the rate of change of the slope is changing. If you're getting a bit lost here, don't worry about it — I'm getting lost myself. It gets increasingly difficult to get a handle on what higher derivatives tell you as you go past the second derivative, because you start getting into a rate of change of a rate of change of a rate of change, and so on.

Chapter 11

Differentiation and the Shape of Curves

· ·

· ·

*I*f you've read Chapters 9 and 10, you're probably an expert at finding derivatives. Which is a good thing, because in this chapter you use derivatives to understand the shape of functions — where they rise and where they fall, where they max out and bottom out, how they curve, and so on. Then in Chapter 12, you use your knowledge about the shape of functions to solve real-world problems.

Taking a Calculus Road Trip

Consider the graph of $f(x)$ in Figure 11-1.

Imagine that you're driving along this function from left to right. Along your drive, there are several points of interest between a and l. All of them, except for the start and finish points, relate to the steepness of the road — in other words, its slope or derivative.

Now, prepare yourself — I'm going to throw lots of new terms and definitions at you all at once here. You shouldn't, however, have much trouble with these ideas because they mostly involve commonsense notions like driving up or down an incline, or going over the crest of a hill.

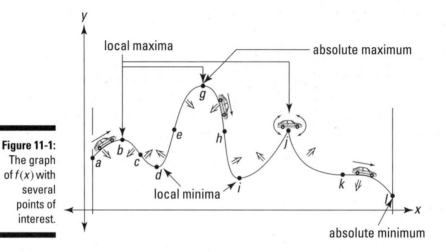

Figure 11-1:
The graph
of $f(x)$ with
several
points of
interest.

Climb every mountain, ford every stream: Positive and negative slopes

First, notice that as you begin your trip at a, you're climbing up. Thus the function is *increasing* and its slope and derivative are therefore *positive*. You climb the hill till you reach the top at b where the road levels out. The road is level there, so the slope and derivative equal *zero*.

Because the derivative is zero at b, point b is called a *stationary point* of the function. Point b is also a *local maximum* or *relative maximum* of f because it's the top of a hill. To be a local max, b just has to be the highest point in its immediate neighborhood. It doesn't matter that the nearby hill at g is even higher.

After reaching the crest of the hill at b, you start going down — duh. So, after b, the slope and derivative are *negative* and the function is *decreasing*. To the left of every local max, the slope is positive; to the right of a max, the slope is negative.

I can't think of a travel metaphor for this section: Concavity and inflection points

The next point of interest is c. Can you see that as you go down from b to c, the road gets steeper and steeper, but that after c, although you're still going down, the road is gradually starting to curve up again and get less steep?

The little down arrow between *b* and *c* in Figure 11-1 indicates that this section of the road is curving down — the function is said to be *concave down* there. As you can see, the road is also concave down between *a* and *b*.

Concavity poetry: *Down* looks like a *frown*, *up* looks like a *cup*. A portion of a function that's concave *down* looks like a *frown*. Where it's concave *up*, like between *c* and *e*, it looks like a *cup*.

Wherever a function is concave *down*, its derivative (and slope) are *decreasing*; wherever a function is concave *up*, its derivative (and slope) are *increasing*.

Okay, so the road is concave down until *c* where it switches to concave up. Because the concavity switches at *c*, it's a *point of inflection*. The point *c* is also the steepest point on this stretch of the road. Inflection points are always at the steepest — or least steep — points in their immediate neighborhoods.

Be careful with function sections that have a negative slope. Point *c* is the steepest point in its neighborhood because it has a bigger negative slope than any other nearby point. But remember, a big negative number is actually a *small* number, so the slope and derivative at *c* are actually the *smallest* of all the points in the neighborhood. From *b* to *c* the derivative of the function is *decreasing* (because it's becoming a bigger negative). From *c* to *d*, the derivative is *increasing* (because it's becoming a smaller negative). Got it?

This vale of tears: A local minimum

Let's get back to your drive. After point *c*, you keep going down till you reach *d*, the bottom of a valley. Point *d* is another stationary point because the road is level there and the derivative is zero. Point *d* is also a *local* or *relative minimum* because it's the lowest point in its immediate neighborhood.

A scenic overlook: The absolute maximum

After *d*, you travel up, passing *e*, which is another inflection point. It's the steepest point between *d* and *g* and the point where the derivative is greatest. You stop at the scenic overlook at *g*, another stationary point and another local max. Point *g* is also the *absolute maximum* on the interval from *a* to *l* because it's the very highest point on the road from *a* to *l*.

Car trouble: Teetering on the corner

Going down from g, you pass another inflection point, h, another local min, i, then you go up to j where you foolishly try to drive over the peak. Your front wheels make it over, but your car's chassis gets stuck on the precipice, leaving you teetering up and down with your wheels spinning. Your car teeters at j because you can't draw a tangent line there. No tangent line means no slope; and no slope means no derivative, or you can say that the derivative at j is *undefined*. A sharp turning point like j is called a *corner*. (By the way, be careful with the expressions "no slope" and "no derivative." In this context, "no" means *nonexistent*, NOT *zero*.)

It's all downhill from here

After dislodging your car, you head down, the road getting less and less steep until it flattens out for an instant at k. (Again, note that because the slope and the derivative are becoming smaller and smaller *negative* numbers on the way to k, they are actually *increasing*.) Point k is another stationary point because its derivative is zero. It's also another inflection point because the concavity switches from up to down at k. After passing k, you go down to l, your final destination. Because l is the endpoint of the interval, it's not a local min — endpoints never qualify as local mins or maxes — but it is the *absolute minimum* on the interval because it's the very lowest point from a to l.

Hope you enjoyed your trip.

Your travel diary

I want to review your trip and some of the previous terms and definitions and introduce yet a few more terms:

- ✔ The function f in Figure 11-1 has a derivative of zero at stationary points (level points) b, d, g, i, and k. At j, the derivative is *undefined*. These points where the derivative is either zero or undefined are the *critical points* of the function. The x-values of these critical points are called the *critical numbers* of the function. (Note that critical numbers must be within a function's domain.)

- ✔ All local maxes and mins — the peaks and valleys — must occur at critical points. However, not all critical points are necessarily local maxes or mins. Point k, for instance, is a critical point but neither a max nor a min. Local maximums and minimums — or *maxima* and *minima* — are called, collectively, local *extrema* of the function. Use a lot of these fancy plurals if you want to sound like a professor. A single local max or min is a local *extremum*. The absolute max is the highest point on the road from *a* to *l*. The absolute min is the lowest point.

✔ The function is increasing whenever you're going up, where the derivative is positive; it's decreasing whenever you're going down, where the derivative is negative. The function is also decreasing at point *k*, a horizontal inflection point, even though the slope and derivative are zero there. I realize that seems a bit odd, but that's the way it works — take my word for it. At all horizontal inflection points, a function is either increasing or decreasing. At local extrema *b*, *d*, *g*, *i*, and *j*, the function is neither increasing nor decreasing.

✔ The function is concave up wherever it looks like a cup or a smile (some say where it "holds water") and concave down wherever it looks like a frown (or "spills water"). Inflection points *c*, *e*, *h*, and *k* are where the concavity switches from up to down or vice versa. Inflection points are also the steepest or least steep points in their immediate neighborhoods.

Finding Local Extrema — My Ma, She's Like, Totally Extreme

Now that you have the preceding section under your belt and know what local extrema are, you need to know how to do the math to find them. You saw in the last section that all local extrema occur at critical points of a function — that's where the derivative is zero or undefined (but don't forget that critical points aren't always local extrema). So, the first step in finding a function's local extrema is to find its critical numbers (the *x*-values of the critical points).

Cranking out the critical numbers

Find the critical numbers of $f(x) = 3x^5 - 20x^3$. See Figure 11-2.

Here's what you do:

1. Find the first derivative of *f* using the power rule.

$$f(x) = 3x^5 - 20x^3$$
$$f'(x) = 15x^4 - 60x^2$$

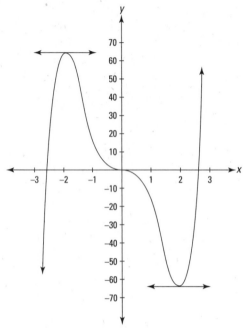

2. Set the derivative equal to zero and solve for x.

$$15x^4 - 60x^2 = 0$$
$$15x^2(x^2 - 4) = 0$$
$$15x^2(x+2)(x-2) = 0$$

$$15x^2 = 0 \quad \text{or} \quad x+2=0 \quad \text{or} \quad x-2=0$$
$$x=0 \quad \text{or} \quad x=-2 \quad \text{or} \quad x=2$$

These three x-values are critical numbers of f. Additional critical numbers could exist if the first derivative were undefined at some x-values, but because the derivative, $15x^4 - 60x^2$, is defined for all input values, the above solution set, 0, –2, and 2, is the complete list of critical numbers. Because the derivative of f equals zero at these three critical numbers, the curve has horizontal tangents at these numbers. In Figure 11-2, you can see the little horizontal tangent lines drawn where $x=-2$ and $x=2$. The third horizontal tangent line where $x=0$ is the x-axis.

A curve has a horizontal tangent line wherever its derivative is zero, namely, at its stationary points. A curve will have horizontal tangent lines at all of its local mins and maxes (except for sharp corners like point j in Figure 11-1) and at all of its horizontal inflection points.

Now that you've got the list of critical numbers, you need to determine whether peaks or valleys or inflection points occur at those *x*-values. You can do this with either the first derivative test or the second derivative test. I suppose you may be wondering why you have to test the critical numbers when you can see where the peaks and valleys are by just looking at the graph in Figure 11-2 — which you can, of course, reproduce on your graphing calculator. Good point. Okay, so this problem — not to mention countless other problems you've done in math courses — is somewhat contrived and impractical. So what else is new?

The first derivative test

The first derivative test is based on the Nobel-Prize-caliber ideas that as you go over the top of a hill, first you go up and then you go down, and that when you drive into and out of a valley, you go down and then up. This calculus stuff is pretty amazing, isn't it?

Here's how you use the test. Take a number line and put down the critical numbers you found above: 0, –2, and 2. See Figure 11-3.

Figure 11-3:
The critical
numbers
of $f(x) =$
$3x^5 - 20x^3$.

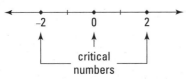

This number line is now divided into four regions: to the left of –2, from –2 to 0, from 0 to 2, and to the right of 2. Pick a value from each region, plug it into the first derivative, and note whether your result is positive or negative. Let's use the numbers –3, –1, 1, and 3 to test the regions:

$$f'(x) = 15x^4 - 60x^2$$

$$f'(-3) = 15(-3)^4 - 60(-3)^2 = 15 \cdot 81 - 60 \cdot 9 = 675$$
$$f'(-1) = 15(-1)^4 - 60(-1)^2 = 15 - 60 = -45$$
$$f'(1) = 15(1)^4 - 60(1)^2 = 15 - 60 = -45$$
$$f'(3) = 15(3)^4 - 60(3)^2 = 15 \cdot 81 - 60 \cdot 9 = 675$$

By the way, if you had noticed that this first derivative is an *even* function, you'd have known, without doing the computation, that $f(1) = f(-1)$ and that $f(3) = f(-3)$. (Chapter 5 discusses even functions. A polynomial function with all even powers, like $f'(x)$ above, is one type of even function.)

These four results are, respectively, positive, negative, negative, and positive. Now, take your number line, mark each region with the appropriate positive or negative sign, and indicate where the function is increasing (where the derivative is positive) and decreasing (where the derivative is negative). The result is a so-called *sign graph*. See Figure 11-4. (The four right-pointing arrows at the top of the figure simply indicate that "increasing" and "decreasing" tell you what's happening as you move along the function from *left to right*.)

Figure 11-4: The sign graph for $f(x) = 3x^5 - 20x^3$.

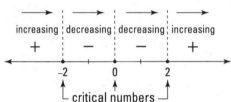

Figure 11-4 simply tells you what you already know if you've looked at the graph of f — that the function goes up until -2, down from -2 to 0, further down from 0 to 2, and up again from 2 on.

Now here's the rocket science. The function switches from increasing to decreasing at -2; in other words, you go up to -2 and then down. So at -2 you have the top of a hill or a local maximum. Conversely, because the function switches from decreasing to increasing at 2, you have the bottom of a valley there or a local minimum. And because the sign of the first derivative doesn't switch (from positive to negative or vice versa) at zero, there's neither a min nor a max at that x-value (you usually — like here — get a horizontal inflection point when this happens).

The last step is to obtain the function values, in other words the heights, of these two local extrema by plugging the x-values into the original function:

$$f(x) = 3x^5 - 20x^3$$

$$f(-2) = 3(-2)^5 - 20(-2)^3 = 64$$
$$f(2) = 3(2)^5 - 20(2)^3 = -64$$

Thus, the local max is located at $(-2, 64)$ and the local min is at $(2, -64)$. You're done.

Local extrema don't occur at points of discontinuity. To use the first derivative test to test for a local extremum at a particular critical number, the function must be *continuous* at that *x*-value.

The second derivative test — no, no, anything but another test!

The second derivative test is based on two more prize-winning ideas: First, that at the crest of a hill, a road has a hump shape — in other words, it's curving down or *concave down;* and second, that at the bottom of a valley, a road is cup-shaped, so it's curving up or *concave up.*

The concavity of a function at a point is given by its second derivative: a *positive* second derivative means the function is concave *up,* a *negative* second derivative means the function is concave *down,* and a second derivative of *zero* is *inconclusive* (the function could be concave up, concave down, or there could be an inflection point there). So, for our function *f,* all you have to do is find its second derivative and then plug in the critical numbers you found, $-2, 0,$ and $2,$ and note whether your results are positive, negative, or zero. To wit —

$$f(x) = 3x^5 - 20x^3$$
$$f'(x) = 15x^4 - 60x^2 \quad \text{(power rule)}$$
$$f''(x) = 60x^3 - 120x \quad \text{(power rule)}$$

$$f''(-2) = 60(-2)^3 - 120(-2) = -240$$
$$f''(0) = 60(0)^3 - 120(0) = 0$$
$$f''(2) = 60(2)^3 - 120(2) = 240$$

At $x = -2,$ the second derivative is negative (-240). This tells you that *f* is concave down where *x* equals $-2,$ and therefore that there's a local max there. The second derivative is positive (240) where *x* is 2, so *f* is concave up and thus there's a local min at $x = 2.$ Because the second derivative equals zero at $x = 0,$ the second derivative test fails for that critical number — it tells you nothing about the concavity at $x = 0$ or whether there's a local min or max there. When this happens, you have to use the first derivative test.

Now go through the first and second derivative tests one more time with another example. Find the local extrema of $g(x) = 2x - 3x^{2/3} + 4.$ See Figure 11-5.

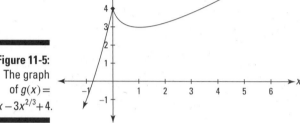

1. **Find the first derivative of g.**

$$g(x)=2x-3x^{2/3}+4$$
$$g'(x)=2-2x^{-1/3} \quad \text{(power rule)}$$

2. **Set the derivative equal to zero and solve.**

$$2-2x^{-1/3}=0$$
$$-2x^{-1/3}=-2$$
$$x^{-1/3}=1$$
$$(x^{-1/3})^{-3}=1^{-3}$$
$$x=1$$

Thus 1 is a critical number.

3. **Determine whether the first derivative is undefined for any x-values.**

$2x^{-1/3}$ equals $\dfrac{2}{\sqrt[3]{x}}$. Now, because the cube root of zero is zero, if you plug in zero to $\dfrac{2}{\sqrt[3]{x}}$, you'd have $\dfrac{2}{0}$, which is undefined. So the derivative, $2-2x^{-1/3}$, is undefined at $x=0$, and thus 0 is another critical number. From Steps 2 and 3, you've got the complete list of critical numbers of g: 0 and 1.

4. **Plot the critical numbers on a number line, and then use the first derivative test to figure out the sign of each region.**

You can use $-1, 0.5,$ and 2 as test numbers:

$$g'(x) = 2 - 2x^{-1/3}$$

$$g'(-1) = 4 \qquad \text{(pos.)}$$
$$g'(0.5) \approx -0.52 \quad \text{(neg.)}$$
$$g'(2) \approx 0.41 \qquad \text{(pos.)}$$

Figure 11-6 shows the sign graph.

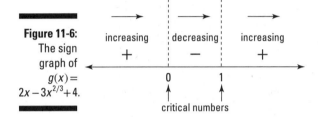

Figure 11-6:
The sign
graph of
$g(x) =$
$2x - 3x^{2/3} + 4.$

increasing   decreasing   increasing

$+$ $-$ $+$

0   1

critical numbers

Because the first derivative of g switches from positive to negative at 0, there's a local max there. And because the first derivative switches from negative to positive at 1, there's a local min at $x = 1$.

5. **Plug the critical numbers into g to obtain the function values (the heights) of these two local extrema.**

$$g(x) = 2x - 3x^{2/3} + 4 \qquad \begin{aligned} g(0) &= 4 \\ g(1) &= 3 \end{aligned}$$

So, there's a local max at $(0, 4)$ and a local min at $(1, 3)$. You're done.

You could have used the second derivative test instead of the first derivative test in Step 4. First, you need the second derivative of g, which is, as you know, the derivative of its first derivative:

$$g'(x) = 2 - 2x^{-1/3}$$
$$g''(x) = \frac{2}{3}x^{-4/3}$$

Evaluate the second derivative at $x = 1$ (the critical number from Step 2):

$$g''(1) = \frac{2}{3}$$

Because $g''(1)$ is positive, you know that g is concave up at $x = 1$ and, therefore, that there's a local min there.

At the other critical number, $x = 0$ (from Step 3), the first derivative is undefined. The second derivative test is no help where the first derivative is undefined, so you've got to use the first derivative test for that critical number.

Finding Absolute Extrema on a Closed Interval

Every function that's continuous on a closed interval has to have an absolute maximum value and an absolute minimum value in that interval — in other words, a highest and lowest point — though, as you see in the following example, there can be a tie for the highest or lowest value.

A *closed* interval like $[2, 5]$ includes the endpoints 2 and 5. An *open* interval like $(2, 5)$ excludes the endpoints.

Finding the absolute max and min is a snap. All you do is compute the critical numbers of the function in the given interval, determine the height of the function at each critical number, and then figure the height of the function at the two endpoints of the interval. The greatest of this set of heights is the absolute max; and the least, of course, is the absolute min. Here's an example: Find the absolute max and min of $h(x) = \cos(2x) - 2\sin x$ in the closed interval $\left[\frac{\pi}{2}, 2\pi\right]$.

1. **Find the critical numbers of h in the *open* interval $\left(\frac{\pi}{2}, 2\pi\right)$.**

   (See Chapter 6 if you're a little rusty on trig functions.)

$$h(x) = \cos(2x) - 2\sin x$$
$$h'(x) = -\sin(2x) \cdot 2 - 2\cos x \quad \text{(by the chain rule)}$$
$$0 = -2\sin(2x) - 2\cos x \quad \text{(now divide both sides by } -2)$$
$$0 = \sin(2x) + \cos x \quad \text{(now use a trig identity)}$$
$$0 = 2\sin x \cos x + \cos x \quad \text{(factor out } \cos x)$$
$$0 = \cos x(2\sin x + 1)$$

$$\cos x = 0 \quad \text{or} \quad 2\sin x + 1 = 0$$
$$x = \frac{3\pi}{2} \qquad \sin x = -\frac{1}{2}$$
$$x = \frac{7\pi}{6}, \frac{11\pi}{6}$$

Thus, the zeros of h' are $\frac{7\pi}{6}, \frac{3\pi}{2}$, and $\frac{11\pi}{6}$, and because h' is defined for all input numbers, this is the complete list of critical numbers.

2. **Compute the function values (the heights) at each critical number.**

$$h(x) = \cos(2x) - 2\sin x$$

$$h\left(\frac{7\pi}{6}\right) = \cos\left(2 \cdot \frac{7\pi}{6}\right) - 2\sin\left(\frac{7\pi}{6}\right)$$
$$= 0.5 - 2 \cdot (-0.5) = 1.5$$

$$h\left(\frac{3\pi}{2}\right) = \cos\left(2 \cdot \frac{3\pi}{2}\right) - 2\sin\left(\frac{3\pi}{2}\right)$$
$$= -1 - 2 \cdot (-1) = 1$$

$$h\left(\frac{11\pi}{6}\right) = \cos\left(2 \cdot \frac{11\pi}{6}\right) - 2\sin\left(\frac{11\pi}{6}\right)$$
$$= 0.5 - 2 \cdot (-0.5) = 1.5$$

3. **Determine the function values at the endpoints of the interval.**

$$h\left(\frac{\pi}{2}\right) = \cos\left(2 \cdot \frac{\pi}{2}\right) - 2\sin\left(\frac{\pi}{2}\right)$$
$$= -1 - 2 \cdot 1 = -3$$
$$h(2\pi) = \cos(2 \cdot 2\pi) - 2\sin(2\pi)$$
$$= 1 - 2 \cdot 0 = 1$$

So, from Steps 2 and 3, you've found five heights: 1.5, 1, 1.5, −3, and 1. The largest number in this list, 1.5, is the absolute max; the smallest, −3, is the absolute min.

The absolute max occurs at two points: $\left(\frac{7\pi}{6}, 1.5\right)$ and $\left(\frac{11\pi}{6}, 1.5\right)$. The absolute min occurs at one of the endpoints, $\left(\frac{\pi}{2}, -3\right)$, and is thus called an *endpoint extremum*.

Table 11-1 shows the values of $h(x) = \cos(2x) - 2\sin x$ at the three critical numbers in the interval from $\frac{\pi}{2}$ to 2π and at the interval's endpoints; Figure 11-7 shows the graph of h.

Table 11-1   **Values of $h(x) = \cos(2x) - 2\sin x$ at the Critical Numbers and Endpoints for the Interval $\left[\frac{\pi}{2}, 2\pi\right]$**

$h(x)$	−3	1.5	1	1.5	1
x	$\frac{\pi}{2}$	$\frac{7\pi}{6}$	$\frac{3\pi}{2}$	$\frac{11\pi}{6}$	2π

A couple observations. First, as you can see in Figure 11-7, the points $\left(\frac{7\pi}{6}, 1.5\right)$ and $\left(\frac{11\pi}{6}, 1.5\right)$ are both *local* maxima of h, and the point $\left(\frac{3\pi}{2}, 1\right)$ is a local minimum of h. However, if you want only to find the *absolute* extrema on a closed interval, you don't have to pay any attention to whether critical points are local maxes, mins, or neither. And thus you don't have to bother to use the first or second derivative tests. All you have to do is determine the heights at the critical numbers and at the endpoints and then pick the largest and smallest numbers from this list. Second, the absolute max and min in the given interval tell you nothing about how the function behaves outside the interval. Function h, for instance, might rise higher than 1.5 outside the interval from $\frac{\pi}{2}$ to 2π (although it doesn't), and it might go lower than -3 (although it never does).

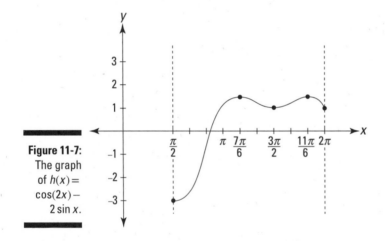

Figure 11-7: The graph of $h(x) = \cos(2x) - 2\sin x$.

Finding Absolute Extrema over a Function's Entire Domain

A function's *absolute max* and *absolute min* over its *entire domain* are the highest and lowest values (heights) of the function anywhere it's defined. Unlike in the previous section where you saw that a continuous function must have both an absolute max and min on a closed interval, when you consider a function's entire domain, a function can have an absolute max or min or both or neither. For example, the parabola $y = x^2$ has an absolute min at the point $(0, 0)$ — the bottom of its cup shape — but no absolute max because it goes up forever to the left and the right. You might think that its absolute max would be infinity, but infinity is not a number and thus it doesn't qualify as a maximum (ditto for using negative infinity as an absolute min).

On the one hand, the idea of a function's very highest point and very lowest point seems pretty simple, doesn't it? But there's a wrench in the works. The wrench is the category of things that *don't* qualify as maxes or mins.

I already mentioned that infinity and negative infinity don't qualify. Then there are empty "endpoints" like $(3, 4)$ on $f(x)$ in Figure 11-8. $f(x)$ doesn't have an absolute max. Its max isn't 4 because it never gets to 4, and its max can't be anything less than 4, like 3.999, because it gets higher than that, say 3.9999. Similarly, an infinitesimal hole in a function can't qualify as a max or min. For example, consider the absolute value function, $y = |x|$, you know, the V-shaped function with the sharp corner at the origin; if you can't picture it, look back at function g in Figure 7-8. $y = |x|$ has no absolute max because it goes up to infinity. Its absolute min is zero (at $(0, 0)$ of course). But now, say you alter the function slightly by plucking out the point at $(0, 0)$ and leaving an infinitesimal hole there. Now the function has no absolute minimum.

Now consider $g(x)$ in Figure 11-8. It shows another type of situation that doesn't qualify as a min (or max). $g(x)$ has no absolute min. Going left, g crawls along the horizontal asymptote at $y = 0$, always getting lower and lower, but never getting as low as zero. Since it never gets to zero, zero can't be the absolute min, and there can't be any other absolute min (like, say, 0.0001) because at some point way to the left, g will get below any small number you can name.

Keeping the above in mind, here's a step-by-step approach for locating a function's absolute maximum and minimum (if there are any):

1. **Find the height of the function at each of its critical numbers. (Recall that a function's critical numbers are the x-values within the function's domain where the derivative is zero or undefined.)**

   You just did this in the previous section, but this time you consider *all* the critical numbers, not just those in a given interval. The highest of these values will be the function's absolute max unless the function goes higher than that point in which case the function won't have an absolute max. The lowest of those values will be the function's absolute min unless the function goes lower than that point in which case it won't have an absolute min. Steps 2 and 3 will help you figure out whether the function goes higher than the highest critical point and/or lower than the lowest critical point. If you apply Step 1 to $g(x)$ in Figure 11-8, you'll find that it has no critical points. When this happens, you're done. The function has neither an absolute max nor an absolute min.

2. **Check whether the function goes up to infinity and/or down to negative infinity.**

   If a function goes up to positive infinity or down to negative infinity, it does so at its extreme right or left or at a vertical asymptote. So, evaluate $\lim_{x \to \infty} f(x)$ and $\lim_{x \to -\infty} f(x)$ — the so-called *end behavior* of the function — and the limit of the function as x approaches each vertical asymptote (if there are any) from the left and from the right. If the function goes up to infinity, it has no absolute max; if it goes down to negative infinity, it has no absolute min.

3. **Graph the function to check for horizontal asymptotes and weird features like the jump discontinuity in $f(x)$ in Figure 11-8.**

   Look at the graph of the function. If you see that the function gets higher than the highest of its critical points, it has no absolute max; if it goes lower than the lowest of its critical points, it has no absolute min. Applying this 3-step process to $f(x)$ in Figure 11-8, Step 1 would reveal two critical points: the endpoint at $(3, 1)$ and the local max at roughly $(4.1, 1.3)$. In Step 2, you would find that f goes down to negative infinity and thus has no absolute min. Finally, in Step 3, you'd see that f goes higher than the higher of the critical points, $(4.1, 1.3)$, and that it, therefore, has no absolute max. You're done!

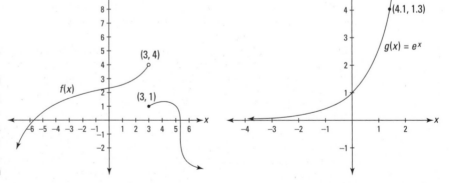

Figure 11-8: Two functions with no absolute extrema.

Locating Concavity and Inflection Points

Look back at the function $f(x) = 3x^5 - 20x^3$ in Figure 11-2. You used the three critical numbers of f, -2, 0, and 2, to find the function's local extrema: $(-2, 64)$ and $(2, -64)$. This section investigates what happens elsewhere on this function — specifically, where it's concave up or down and where the concavity switches (the inflection points).

The process for finding concavity and inflection points is analogous to using the first derivative test and the sign graph to find local extrema, except that now you use the second derivative. (See the section "Finding Local Extrema") Here's what you do to find the intervals of concavity and the inflection points of $f(x) = 3x^5 - 20x^3$:

1. **Find the second derivative of *f*.**

   $$f(x) = 3x^5 - 20x^3$$
   $$f' = 15x^4 - 60x^2 \quad \text{(the power rule)}$$
   $$f'' = 60x^3 - 120x \quad \text{(the power rule)}$$

2. **Set the second derivative equal to zero and solve.**

   $$60x^3 - 120x = 0$$
   $$60x(x^2 - 2) = 0$$

   $$60x = 0 \qquad \text{or} \qquad x^2 - 2 = 0$$
   $$x = 0 \qquad\qquad\qquad x^2 = 2$$
   $$x = \pm\sqrt{2}$$

3. **Determine whether the second derivative is undefined for any *x*-values.**

   $f'' = 60x^3 - 120x$ is defined for all real numbers, so there are no other x-values to add to the list from Step 2. Thus, the complete list is $-\sqrt{2}$, 0, and $\sqrt{2}$.

   Steps 2 and 3 give you what you could call "second derivative critical numbers" of f because they're analogous to the critical numbers of f that you find using the first derivative. But, as far as I'm aware, this set of numbers has no special name. The important thing to know is that this list is made up of the zeros of f'' plus any x-values where f'' is undefined.

4. Plot these numbers on a number line and test the regions with the *second* derivative.

Use –2, –1, 1, and 2 as test numbers.

$$f''(x) = 60x^3 - 120x$$

$$f''(-2) = -240 \quad \text{(neg.)}$$
$$f''(-1) = 60 \quad \text{(pos.)}$$
$$f''(1) = -60 \quad \text{(neg.)}$$
$$f''(2) = 240 \quad \text{(pos.)}$$

Figure 11-9 shows the sign graph.

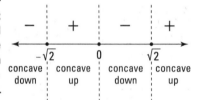

Figure 11-9:
A second derivative sign graph for $f(x) = 3x^5 - 20x^3$.

A positive sign on this sign graph tells you that the function is concave up in that interval; negative means concave down. The function has an inflection point (usually) at any x-value where the signs switch from positive to negative or vice versa.

Because the signs switch at $-\sqrt{2}$, 0, and $\sqrt{2}$, and because these three numbers are zeros of f'', inflection points occur at these x-values. If, however, you have a problem where the signs switch at a number where f'' is undefined, you have to check one additional thing before concluding that there's an inflection point there. An inflection point exists at a given x-value only if you can draw a tangent line to the function at that number. This is the case if the first derivative exists at that number or if the tangent line is vertical there.

5. Plug these three x-values into f to obtain the function values of the three inflection points.

$$f(x) = 3x^5 - 20x^3$$

$$f(-\sqrt{2}) \approx -39.6$$
$$f(0) = 0$$
$$f(\sqrt{2}) \approx -39.6$$

The square root of two equals about 1.4, so there are inflection points at about $(-1.4, 39.6)$, $(0, 0)$, and about $(1.4, -39.6)$. You're done.

Figure 11-10 shows f's inflection points as well as its local extrema and its intervals of concavity.

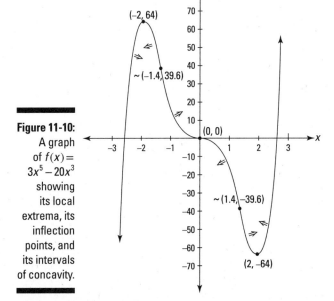

Figure 11-10:
A graph of $f(x) = 3x^5 - 20x^3$ showing its local extrema, its inflection points, and its intervals of concavity.

Looking at Graphs of Derivatives Till They Derive You Crazy

You can learn a lot about functions and their derivatives by looking at their graphs side by side and comparing their important features. Let's keep going with the same function, $f(x) = 3x^5 - 20x^3$; we're going to travel along f from left to right (see Figure 11-11), pausing to note its points of interest and also observing what's happening to the graph of $f' = 15x^4 - 60x^2$ at the same points. But first, check out the (long) warning beneath the figure.

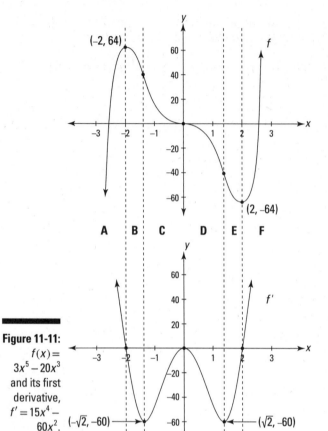

Figure 11-11:
$f(x) = 3x^5 - 20x^3$ and its first derivative, $f' = 15x^4 - 60x^2$.

WARNING!

This is NOT the function! As you look at the graph of f' in Figure 11-11, or the graph of any other derivative, you may need to slap yourself in the face every minute or so to remind yourself that "This is the *derivative* I'm looking at, *not* the function! You've looked at hundreds and hundreds of graphs of functions over the years, so when you start looking at graphs of derivatives, you can easily lapse into thinking of them as regular functions. You might, for instance, look at an interval that's going up on the graph of a derivative and mistakenly conclude that the original function must also be going up in the same interval — an understandable mistake. You know the first derivative is the same thing as slope. So when you see the graph of the first derivative going up, you may think, "Oh, the first derivative (the slope) is going up, and when the slope goes up that's like going up a hill, so the original function must be rising." This sounds reasonable because, *loosely* speaking, you can describe the front side of a hill as a slope that's going up, increasing. But *mathematically* speaking, the front side of a hill has a *positive* slope, not necessarily an *increasing* slope. So, where a function is increasing, the graph of its derivative will be *positive*, but that derivative graph might be going up or down. Say you're going up a hill. As you approach the top of the hill, you're still going *up,* but, in general, the *slope* (the steepness) is going *down.* It might be 3, then 2, then 1, and then, at the top of the hill, the slope is zero. So the slope is getting smaller or *decreasing,* even as you're climbing the hill or *increasing.* In such an interval, the graph of the function is *increasing,* but the graph of its derivative is *decreasing.* Got that?

Okay, let's get back to the *f* and its derivative in Figure 11-11. Beginning on the left and traveling toward the right, *f* increases until the local max at $(-2, 64)$. It's going up, so its slope is *positive,* but *f* is getting less and less steep so its slope is *decreasing* — the slope decreases until it becomes zero at the peak. This corresponds to the graph of *f′* (the slope) which is *positive* (because it's above the *x*-axis) but *decreasing* as it goes down to the point $(-2, 0)$. Let's summarize your entire trip along *f* and *f′* with the following list of rules.

Rules are rules:

✔ An *increasing* interval on a function corresponds to an interval on the graph of its derivative that's *positive* (or *zero* for a single point if the function has a horizontal inflection point). In other words, a function's increasing interval corresponds to a part of the derivative graph that's above the *x*-axis (or that touches the axis for a single point in the case of a horizontal inflection point). See intervals A and F in Figure 11-11.

✔ A local *max* on the graph of a function (like $(-2, 64)$ corresponds to a *zero* (an *x*-intercept) on an interval of the graph of its derivative that crosses the *x*-axis going *down* (like at $(-2, 0)$).

On a derivative graph, you've got an *m*-axis. When you're looking at various points on the derivative graph, don't forget that the *y*-coordinate of a point, like $(-2, 0)$, on a graph of a first derivative tells you the *slope* of the original function, not its height. Think of the *y*-axis on the first derivative graph as the *slope*-axis or the *m*-axis; you could think of general points on the first derivative graph as having coordinates (x, m).

✔ A *decreasing* interval on a function corresponds to a *negative* interval on the graph of the derivative (or *zero* for a single point if the function has a horizontal inflection point). The negative interval on the derivative graph is below the *x*-axis (or in the case of a horizontal inflection point, the derivative graph touches the *x*-axis at a single point). See intervals B, C, D, and E in Figure 11-11 (but consider them as a single section), where *f* goes down all the way from the local max at $(-2, 64)$ to the local min at $(-2, 64)$ and where *f′* is negative between $(-2, 0)$ and $(2, 0)$ except for at the point $(0, 0)$ on *f′* which corresponds to the horizontal inflection point on *f*.

✔ A local *min* on the graph of a function corresponds to a zero (an *x*-intercept) on an interval of the graph of its derivative that crosses the *x*-axis going up (like at $(2, 0)$).

Now let's take a second trip along *f* to consider its intervals of concavity and its inflection points. First, consider intervals A and B in Figure 11-11. The graph of *f* is concave down — which means the same thing as a *decreasing* slope — until it gets to the inflection point at about (–1.4, 39.6).

So, the graph of *f′* *decreases* until it bottoms out at about (–1.4, –60). These coordinates tell you that the inflection point at –1.4 on *f* has a slope of –60. Note that the inflection point on *f* at (–1.4, 39.6) is the steepest point on that stretch of the function, but it has the *smallest* slope because its slope is a larger *negative* than the slope at any other nearby point.

Between (–1.4, 39.6) and the next inflection point at (0, 0), *f* is concave up, which means the same thing as an *increasing* slope. So the graph of *f′* *increases* from about –1.4 to where it hits a local max at (0, 0). See interval C in Figure 11-11. Let's take a break from our trip for some more rules.

More rules:

✔ A concave *down* interval on the graph of a function corresponds to a *decreasing* interval on the graph of its derivative (intervals A, B, and D in Figure 11-11). And a concave *up* interval on the function corresponds to an *increasing* interval on the derivative (intervals C, E, and F).

✔ An *inflection point* on a function (except for a vertical inflection point where the derivative is undefined) corresponds to a *local extremum* on the graph of its derivative. An inflection point of *minimum* slope (in its neighborhood) corresponds to a local *min* on the derivative graph; an inflection point of *maximum* slope (in its neighborhood) corresponds to a local *max* on the derivative graph.

Resuming our trip, after (0, 0), *f* is concave down till the inflection point at about (–1.4, 39.6) — this corresponds to the decreasing section of *f′* from (0, 0) to its min at (1.4, –60) (interval D in Figure 11-11). Finally, *f* is concave up the rest of the way, which corresponds to the increasing section of *f′* beginning at (1.4, –60) (intervals E and F in the figure).

Well, that pretty much brings you to the end of the road. Going back and forth between the graphs of a function and its derivative can be *very* trying at first. If your head starts to spin, take a break and come back to this stuff later.

If I haven't already succeeded in *deriving* you crazy — aren't these calculus puns fantastic? — perhaps this final point will do the trick. Look again at the graph of the derivative, *f′*, in Figure 11-11 and also at the sign graph for *f′* in Figure 11-9. That sign graph, because it's a second derivative sign graph, bears exactly (well, almost exactly) the same relationship to the graph of *f′* as a first derivative sign graph bears to the graph of a regular function. In other words, *negative* intervals on the sign graph in Figure 11-9 (to the left of $-\sqrt{2}$ and between 0 and $\sqrt{2}$) show you where the graph of *f′* is *decreasing*; *positive* intervals on the sign graph (between $-\sqrt{2}$ and 0 and to the right of $\sqrt{2}$) show you where f' is *increasing*. And points where the signs switch from positive to negative or vice versa (at $-\sqrt{2}$, 0, and $\sqrt{2}$) show you where f' has local extrema. Clear as mud, right?

The Mean Value Theorem — GRRRRR

You won't use the mean value theorem a lot, but it's a famous theorem — one of the two or three most important in all of calculus — so you really should learn it. It's very simple and has a nice connection to the mean value theorem for integrals which I show you in Chapter 17. Look at Figure 11-12.

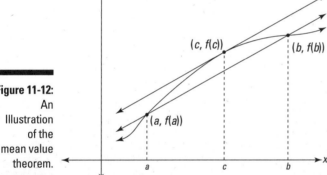

Figure 11-12:
An
Illustration
of the
mean value
theorem.

Here's the formal definition of the theorem.

The mean value theorem: If f is continuous on the closed interval $[a, b]$ and differentiable on the open interval (a, b) then there exists at least one number c in (a, b) such that

$$f'(c) = \frac{f(b) - f(a)}{b - a}$$

Now for the plain-English version. First you need to take care of the fine print. The requirements in the theorem that the function be continuous and differentiable just guarantee that the function is a regular, smooth function without gaps or corners. But because only a few weird functions have gaps or corners, you don't often have to worry about these fine points.

Here's what the theorem means. The secant line connecting points $(a, f(a))$ and $(b, f(b))$ in Figure 11-12 has a slope given by the slope formula:

$$Slope = \frac{y_2 - y_1}{x_2 - x_1}$$
$$= \frac{f(b) - f(a)}{b - a}$$

Note that this is the same as the right side of the equation in the mean value theorem. The derivative at a point is the same thing as the slope of the tangent line at that point, so the theorem just says that there must be at least one point between a and b where the slope of the tangent is the same as the slope of the secant line from a to b. The result is parallel lines, like in Figure 11-12.

Why must this be so? Here's a visual argument. Imagine that you grab the secant line connecting $(a, f(a))$ and $(b, f(b))$, and then you slide it up, keeping it parallel to the original secant line. Can you see that the two points of intersection between this sliding line and the function — the two points that begin at $(a, f(a))$ and $(b, f(b))$ — will gradually get closer and closer to each other until they come together at $(c, f(c))$? If you raise the line any further, you break away from the function entirely. At this last point of intersection, $(c, f(c))$, the sliding line touches the function at a single point and is thus tangent to the function there, and it has the same slope as the original secant line. Well, that does it. This explanation is a bit oversimplified, but it'll do.

Here's a completely different sort of argument that should appeal to your common sense. If the function in Figure 11-12 gives your car's odometer reading as a function of time, then the slope of the secant line from a to b gives your average speed during that interval of time, because dividing the distance traveled, $f(b) - f(a)$, by the elapsed time, $b - a$, gives you the average speed. The point $(c, f(c))$, guaranteed by the mean value theorem, is a point where your instantaneous speed — given by the derivative $f'(c)$ — equals your average speed.

Now, imagine that you take a drive and average 50 miles per hour. The mean value theorem guarantees that you are going exactly 50 mph for at least one moment during your drive. Think about it. Your average speed can't be 50 mph if you go slower than 50 the whole way or if you go faster than 50 the whole way. To average 50 mph, either you go exactly 50 for the whole drive, or you have to go slower than 50 for part of the drive and faster than 50 at other times. And if you're going less than 50 at one point and more than 50 at a later point (or vice versa), you've got to hit exactly 50 at least once as you speed up (or slow down). You can't jump over 50 — like going 49 one moment then 51 the next — because speeds go up by *sliding* up the scale, not jumping. At some point your speedometer slides past 50, and for at least one instant, you're going exactly 50 mph. That's all the mean value theorem says.

Chapter 12

Your Problems Are Solved: Differentiation to the Rescue!

In This Chapter

▶ Getting the most bang for your buck — optimization problems

▶ Position, velocity, and acceleration — VROOOOM

▶ Related rates — brace yourself

*I*n the Introduction, I argue that calculus has changed the world in countless ways, that its impact is not limited to Ivory Tower mathematics, but is all around us in down-to-earth things like microwave ovens, cell phones, and cars. Well, it's now Chapter 12, and I'm *finally* ready to show you how to use calculus to solve some practical problems.

Getting the Most (or Least) Out of Life: Optimization Problems

One of the most practical uses of differentiation is finding the maximum or minimum value of a real-world function: the maximum output of a factory, the maximum strength of a beam, the minimum time to accomplish some task, the maximum range of a missile, and so on. Let's get started with a couple standard geometry examples.

The maximum volume of a box

A box with no top is to be manufactured from a 30-inch-by-30-inch piece of cardboard by cutting and folding it, as shown in Figure 12-1.

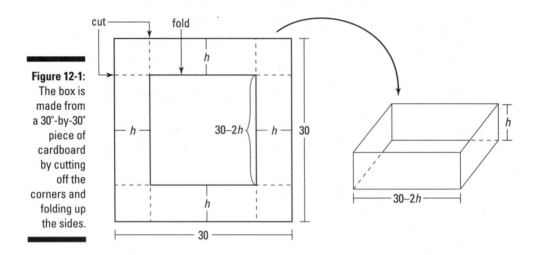

Figure 12-1:
The box is made from a 30"-by-30" piece of cardboard by cutting off the corners and folding up the sides.

What dimensions will produce a box with the maximum volume? Mathematics often seems abstract and impractical, but here's an honest-to-goodness practical problem (well . . . almost). If a manufacturer can sell bigger boxes for more and is making 100,000 boxes, you'd better believe he or she wants the exact answer to this question. Here's how you do it:

1. **Express the thing you want maximized, the volume, as a function of the unknown, the height of the box (which is the same as the length of the cut).**

$$V = l \cdot w \cdot h$$

$$V(h) = (30 - 2h)(30 - 2h) \cdot h$$ (You can see in Figure 12-1 that both the *length* and *width* equal $30 - 2h$.)

$$= (900 - 120h + 4h^2) \cdot h$$
$$= 4h^3 - 120h^2 + 900h$$

2. **Determine the domain of your function.**

The height can't be negative, and because the length (and width) of the box equals $30 - 2h$, which can't be negative, h can't be greater than 15. Thus, sensible values for h are $0 \le h \le 15$. You now want to find the maximum value of $V(h)$ in this interval. You use the method from the "Finding Absolute Extrema on a Closed Interval" section in Chapter 11.

3. **Find the critical numbers of $V(h)$ in the open interval $(0, 15)$ by setting its derivative equal to zero and solving. And don't forget to check for numbers where the derivative is undefined.**

$$V(h) = 4h^3 - 120h^2 + 900h$$
$$V'(h) = 12h^2 - 240h + 900 \quad \text{(power rule)}$$
$$0 = 12h^2 - 240h + 900$$
$$0 = h^2 - 20h + 75 \quad \text{(dividing both sides by 12)}$$
$$h = (h - 15)(h - 5) \quad \text{(ordinary trinomial factoring)}$$
$$h = 15 \text{ or } 5$$

Because 15 is not in the open interval $(0, 15)$, it doesn't qualify as a critical number (though this is a moot point because you end up testing it in Step 4 anyway). And because this derivative is defined for all input values and is, thus, of course, never *undefined*, there are no additional critical numbers. So 5 is the only critical number.

4. **Evaluate the function at the critical number, 5, and at the endpoints of the interval, 0 and 15, to locate the function's max.**

$$V(h) = 4h^3 - 120h^2 + 900h$$

$$V(0) = 0$$
$$V(5) = 2000$$
$$V(15) = 0$$

Test the endpoints. The *extremum* (dig that fancy word for *maximum* or *minimum*) you're looking for doesn't often occur at an endpoint, but it can — so don't fail to evaluate the function at the interval's two endpoints.

So, a height of 5 inches produces the box with maximum volume (2,000 cubic inches). Because the length and width equal $30 - 2h$, a height of 5 gives a length and width of $30 - 2 \cdot 5$, or 20, and thus the dimensions of the desired box are $5'' \times 20'' \times 20''$. That's it.

The maximum area of a corral — yeehaw!

A rancher can afford 300 feet of fencing to build a corral that's divided into two equal rectangles. See Figure 12-2.

What dimensions will maximize the corral's area? This is another practical problem. The rancher wants to give his animals as much room as possible given the length of fencing he can afford. Like all businesspeople, he wants the most bang for his buck.

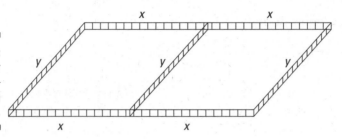

Figure 12-2:
Calculus for
cowboys —
maximizing
a corral.

1.a. Express the thing you want maximized (area) as a function of the two unknowns (x and y).

$$A = l \cdot w$$
$$= (2x)(y)$$

In the cardboard box example in the previous section, you can easily write the volume as a function of *one* variable — which is always what you need. But here, the area is a function of two variables (x and y), so Step 1 has the following two extra sub-steps that will eliminate one of the variables.

1.b. Use the given information to relate the two variables to each other.

The 300 feet of fencing is used for seven sections, thus

$$300 = x + x + x + x + y + y + y$$
$$300 = 4x + 3y$$

1.c. Solve this equation for y and plug the result in for y in the equation from Step 1.a. This gives you what you need — a function of one variable.

$$4x + 3y = 300$$
$$3y = 300 - 4x$$
$$y = \frac{300 - 4x}{3}$$
$$y = 100 - \frac{4}{3}x \quad \text{(Now do the substitution.)}$$

$$A = (2x)(y)$$
$$A(x) = (2x)\left(100 - \frac{4}{3}x\right)$$
$$A(x) = 200x - \frac{8}{3}x^2$$

2. Determine the domain of the function.

You can't have a negative length of fence, so x can't be negative, and the most x can be is 300 divided by 4, or 75. Thus, $0 \le x \le 75$.

3. Find the critical numbers of $A(x)$ in the open interval $(0, 75)$ by setting its derivative equal to zero and solving (and check whether the derivative is undefined anywhere in the interval).

$$A(x) = 200x - \frac{8}{3}x^2$$

$$A'(x) = 200 - \frac{16}{3}x \quad \text{(power rule)}$$

$$0 = 200 - \frac{16}{3}x$$

$$\frac{16}{3}x = 200$$

$$x = 200 \cdot \frac{3}{16}$$

$$= 37.5$$

Because A' is defined for all x-values, 37.5 is the only critical number.

4. Evaluate the function at the critical number, 37.5, and at the endpoints of the interval, 0 and 75.

$$A(x) = 200x - \frac{8}{3}x^2$$

$$A(0) = 0$$
$$A(37.5) = 3750$$
$$A(75) = 0$$

Note: Evaluating a function at the endpoints of a closed interval is a standard step in finding an absolute extremum on the interval. However, you could have skipped this step here had you noticed that $A(x)$ is an upside-down parabola and that, therefore, its peak at $(37.5, 3750)$ must be higher than either endpoint.

The maximum value in the interval is 3,750, and thus, an x-value of 37.5 feet maximizes the corral's area. The length is $2x$, or 75 feet. The width is y, which equals $100 - \frac{4}{3}x$. Plugging in 37.5 gives you $100 - \frac{4}{3}(37.5)$, or 50 feet. So the rancher will build a $75' \times 50'$ corral with an area of 3,750 square feet.

Yo-Yo a Go-Go: Position, Velocity, and Acceleration

Every time you get in your car, you witness differentiation. Your speed is the first derivative of your position. And when you step on the accelerator or the brake — accelerating or decelerating — you experience a second derivative.

The derivative of position is velocity, and the derivative of velocity is acceleration. If a function gives the position of something as a function of time, you differentiate the *position* function to get the *velocity* function, and you differentiate the *velocity* function to get the *acceleration* function. Stated a different but equivalent way, the first derivative of position is velocity, and the second derivative of position is acceleration.

Here's an example. A yo-yo moves straight up and down. Its height above the ground, as a function of time, is given by the function $H(t) = t^3 - 6t^2 + 5t + 30$, where t is in seconds and $H(t)$ is in inches. At $t = 0$, it's 30 inches above the ground, and after 4 seconds, it's at a height of 18 inches. See Figure 12-3.

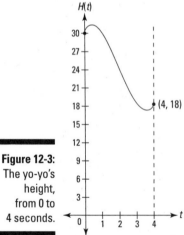

Figure 12-3:
The yo-yo's height, from 0 to 4 seconds.

Velocity, $V(t)$, is the derivative of position (height, in this problem), and acceleration, $A(t)$, is the derivative of velocity. Thus:

$$H(t) = t^3 - 6t^2 + 5t + 30$$

$$V(t) = H'(t) = 3t^2 - 12t + 5 \quad \text{(power rule)}$$

$$A(t) = V'(t) = H''(t) = 6t - 12 \quad \text{(power rule)}$$

Take a look at the graphs of these three functions in Figure 12-4.

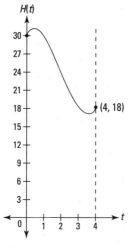

$$H(t) = t^3 - 6t^2 + 5t + 30$$

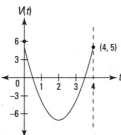

$$V(t) = 3t^2 - 12t + 5$$

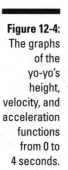

Figure 12-4:
The graphs
of the
yo-yo's
height,
velocity, and
acceleration
functions
from 0 to
4 seconds.

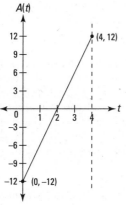

$$A(t) = 6t - 12$$

Using the three functions and their graphs, I want to discuss several things about the yo-yo's motion:

- ✔ Maximum and minimum height
- ✔ Maximum, minimum, and average velocity
- ✔ Total displacement
- ✔ Maximum, minimum, and average speed
- ✔ Total distance traveled
- ✔ Positive and negative acceleration
- ✔ Speeding up and slowing down

Because this is a lot to cover, I'll cut some corners — like not always checking endpoints when looking for extrema if it's obvious that they don't occur at the endpoints. Do you mind? I didn't think so. (Position, velocity, and acceleration problems make use of several ideas from Chapter 11 — local extrema, concavity, inflection points — so you may want to take a look back at those definitions if you're a little hazy.) But before tackling the bulleted topics, let's go over a few things about velocity, speed, and, acceleration.

Velocity, speed, and acceleration

None of your friends will complain — or even notice — if you use the words "velocity" and "speed" interchangeably, but your friendly mathematician *will* complain. Here's the difference. For the velocity function in Figure 12-4, *upward* motion by the yo-yo is defined as a *positive* velocity, and *downward* motion is a *negative* velocity. This is the standard way velocity is treated in most calculus and physics problems. (Or, if the motion is horizontal, going *right* is a *positive* velocity and going *left* is a *negative* velocity.)

Speed, on the other hand, is always positive (or zero). If a car goes by at 50 mph, for instance, you say its speed is 50, and you mean *positive* 50, regardless of whether it's going to the right or the left. For velocity, the direction matters; for speed, it does not. In everyday life, speed is a simpler idea than velocity because it's always positive and because it agrees with our commonsense notion about how fast something is moving. But in calculus, speed is actually the trickier idea because it doesn't fit nicely into the three-function scheme shown in Figure 12-4.

You've got to keep the velocity-speed distinction in mind when analyzing velocity and acceleration. The way we talk about velocity, speed, and acceleration — in calculus class, as opposed to in everyday life — can get pretty weird. For example, if an object is going down (or to the left) faster and faster, its speed is increasing, but its velocity is *decreasing* because its velocity is becoming a bigger and bigger *negative* (and bigger negatives are smaller numbers). This seems weird, but that's the way it works. And here's another strange thing: Acceleration is defined as the rate of change of velocity, not speed. So, if an object is slowing down while going in the downward direction and thus has an *increasing* velocity — because the velocity is becoming a smaller and smaller negative — the object has a *positive* acceleration. In everyday English, you'd say that the object is decelerating (slowing down), but in calculus class, though you could still say the object is slowing down, you'd say that the object has a negative velocity and a positive acceleration. (By the way, *deceleration* isn't exactly a technical term, so you should probably avoid it in calculus class. It's best to use the following vocabulary: *positive acceleration, negative acceleration, positive velocity, negative velocity, speeding up,* and *slowing down.*) I could go on with this stuff, but I bet you've had enough.

Maximum and minimum height

The maximum and minimum height of the yo-yo, in other words, the max and min of $H(t)$, occur at the local extrema you can see in Figure 12-4. To locate them, set the derivative of $H(t)$ (that's $V(t)$) equal to zero and solve:

$$H'(t) = V(t) = 3t^2 - 12t + 5$$
$$0 = 3t^2 - 12t + 5$$
$$t = \frac{-(-12) \pm \sqrt{(-12)^2 - 4(3)(5)}}{2 \cdot 3} \quad \text{(quadratic formula)}$$
$$= \frac{12 \pm \sqrt{84}}{6}$$
$$= \frac{12 \pm 2\sqrt{21}}{6}$$
$$= \frac{6 \pm \sqrt{21}}{3}$$
$$= {\sim}0.47 \text{ or } {\sim}3.53$$

These two numbers are the zeros of $H'(t)$ (which is $V(t)$) and the t-coordinates, that's *time*-coordinates, of the max and min of $H(t)$, which you can see in Figure 12-4. In other words, these are the *times* when the yo-yo reaches its maximum and minimum heights. Plug these numbers into $H(t)$ to obtain the heights:

$$H(0.47) \approx 31.1$$
$$H(3.53) \approx 16.9$$

So the yo-yo gets as high as about 31.1 inches above the ground at $t \approx 0.47$ seconds and as low as about 16.9 inches at $t \approx 3.53$ seconds. (By the way, do you see why the max and min of the yo-yo's height would occur when the yo-yo's velocity is zero?)

Velocity and displacement

As I explain in the "Velocity versus speed" section, *velocity* is basically like *speed* except that while speed is always positive (or zero), velocity can be positive (when going up or to the right) or negative (when going down or to the left). The connection between *displacement* and *distance traveled* is similar: Distance traveled is always positive (or zero), but going down (or left) counts as *negative* displacement. In everyday speech, speed and distance traveled are the more user-friendly ideas, but when it comes to calculus and physics, velocity and displacement are the more fundamental ideas.

Total displacement

Let's get back to our yo-yo analysis. Total displacement is defined as final position minus initial position. So, because the yo-yo starts at a height of 30 and ends at a height of 18,

$$Total\ displacement = 18 - 30 = -12\ inches$$

This is negative because the net movement is *downward*.

Average velocity

Average velocity is given by total displacement divided by elapsed time. We just calculated the total distance (-12 inches), and time runs from 0 seconds to 4 seconds, so the elapsed time is 4 seconds. Thus,

$$Average\ velocity = \frac{-12\ inches}{4\ seconds} = -3\ inches\ per\ second$$

This answer of *negative* 3 tells you that the yo-yo is, on average, going *down* 3 inches per second.

Maximum and minimum velocity

To determine the yo-yo's maximum and minimum velocity during the interval from 0 to 4 seconds, set the derivative of $V(t)$ (that's $A(t)$) equal to zero and solve:

$$V'(t) = A(t) = 6t - 12$$
$$6t - 12 = 0$$
$$6t = 12$$
$$t = 2$$

(Look again at Figure 12-4. At $t = 2$, you get the zero of $A(t)$, the local min of $V(t)$, and the inflection point of $H(t)$. But you already knew that, right? If not, check out Chapter 11.)

Now, evaluate $V(t)$, at the critical number, 2, and at the interval's endpoints, 0 and 4:

$$V(t) = 3t^2 - 12t + 5$$

$$V(0) = 5$$
$$V(2) = -7$$
$$V(4) = 5$$

So, the yo-yo has a maximum velocity of 5 inches per second twice — at both the beginning and the end of the interval. It reaches a minimum velocity of -7 inches per second at $t = 2$ seconds.

Speed and distance traveled

As mentioned in the previous section, *velocity* and *displacement* are the more technical concepts, while *speed* and *distance traveled* are the more common-sense ideas. *Speed,* of course, is the thing you read on your speedometer, and you can read *distance traveled* on your odometer or your "tripometer" after setting it to zero.

Total distance traveled

To determine total distance, add up the distances traveled on each leg of the yo-yo's trip: the up leg, the down leg, and the second up leg.

First, the yo-yo goes up from a height of 30 inches to about 31.1 inches (where the first turn-around point is). That's a distance of about 1.1 inches. Next, it goes down from about 31.1 to about 16.9 (the height of the second turn-around point). That's a distance of 31.1 minus 16.9, or about 14.2 inches. Finally, the yo-yo goes up again from about 16.9 inches to its final height of 18 inches. That's another 1.1 inches. Add these three distances to obtain the total distance traveled: $\sim 1.1 + \sim 14.2 + \sim 1.1 \approx 16.4$ inches. (***Note:*** Compare this answer to the total displacement of –12. The displacement is negative because the net movement is downward. And the positive amount of the displacement (namely 12) is less than the distance traveled of 16.4 because with displacement the up legs of the yo-yo's trip cancel out part of the down leg distance. Check out the math: $1.1 - 14.2 + 1.1 = -12$. Get it?)

Average speed

The yo-yo's average speed is given by the total distance traveled divided by the elapsed time. Thus,

$$Average\ speed \approx \frac{16.4}{4} \approx 4.1\ inches\ per\ second$$

Maximum and minimum speed

You previously determined the yo-yo's maximum velocity (5 inches per second) and its minimum velocity (–7 inches per second). A velocity of –7 is a speed of 7, so that's the yo-yo's maximum speed. Its minimum speed of zero occurs at the two turnaround points.

A good way to analyze maximum and minimum speed is to consider the speed function and its graph. (Or, if you're a glutton for punishment, check out the mumbo jumbo below.) Speed equals the absolute value of velocity. So, for our yo-yo problem, the speed function, $S(t)$, equals $|V(t)| = |3t^2 - 12t + 5|$.

Check out the graph of $S(t)$ in Figure 12-5. Looking at this graph, it's easy to see that the yo-yo's maximum speed occurs at $t = 2$ (the maximum speed is $S(2) = |3(2)^2 - 12(2) + 5| = 7$) and that the minimum speed is zero at the two x-intercepts.

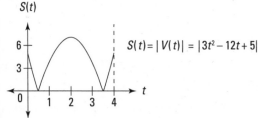

Figure 12-5: The yo-yo's speed function, $S(t) = |V(t)|$.

$$S(t) = |V(t)| = |3t^2 - 12t + 5|$$

Minimum and maximum speed: For a continuous *velocity* function, the *minimum speed* is zero whenever the maximum and minimum velocities are of opposite signs or when one of them is zero. When the maximum and minimum velocities are both positive or both negative, then the *minimum* speed is the *lesser* of the absolute values of the maximum and minimum velocities. In all cases, the *maximum* speed is the *greater* of the absolute values of the maximum and minimum velocities. Is that a mouthful or what?

Burning some rubber with acceleration

Let's go over acceleration: Put your pedal to the metal.

Positive and negative acceleration

The graph of the acceleration function at the bottom of Figure 12-4 is a simple line, $A(t) = 6t - 12$. It's easy to see that the acceleration of the yo-yo goes from a minimum of $-12 \frac{\text{inches per second}}{\text{second}}$ at $t = 0$ seconds to a maximum of $12 \frac{\text{inches per second}}{\text{second}}$ at $t = 4$ seconds, and that the acceleration is zero at $t = 2$ when the yo-yo reaches its minimum velocity (and maximum speed). When the acceleration is *negative* — on the interval $[0, 2)$ — that means that the velocity is *decreasing*. When the acceleration is *positive* — on the interval $(2, 4]$ — the velocity is *increasing*.

Speeding up and slowing down

Figuring out when the yo-yo is speeding up and slowing down is probably more interesting and descriptive of its motion than the info in the preceding section. An object is speeding up (what we call "acceleration" in everyday speech) whenever the velocity and the calculus acceleration are both positive or both negative. And an object is slowing down ("deceleration" in everyday speech) when the velocity and the calculus acceleration are of opposite signs.

Look at all three graphs in Figure 12-4 again. From $t = 0$ to about $t = 0.47$, the velocity is positive and the acceleration is negative, so the yo-yo is slowing down while moving upward (till its velocity becomes zero and it reaches its maximum height). In plain English, the yo-yo is decelerating from 0 to about 0.47 seconds. The greatest deceleration occurs at $t = 0$ when the deceleration is $12 \frac{\text{inches per second}}{\text{second}}$ (the graph shows *negative* 12, but I'm calling it positive 12 because I'm calling it a deceleration, get it?)

From about $t = 0.47$ to $t = 2$, both velocity and acceleration are negative, so the yo-yo is speeding up while moving downward. From $t = 2$ to about $t = 3.53$, velocity is negative and acceleration is positive, so the yo-yo is slowing down again as it continues downward (till it bottoms out at its lowest height). Finally, from about $t = 3.53$ to $t = 4$, both velocity and acceleration are positive, so the yo-yo is speeding up again. The yo-yo reaches its greatest acceleration of $12 \frac{\text{inches per second}}{\text{second}}$ at $t = 4$ seconds.

What the heck is a second squared?

Note that I use the unit $\frac{\text{inches per second}}{\text{second}}$ for acceleration instead of the equivalent but weird-looking unit, $\frac{\text{inches}}{\text{second}^2}$. You often see acceleration given in terms of a unit of distance over second^2, or you might see something like inches per second^2. But what the heck is a second^2? It's meaningless, and something like inches/second^2 is a bad way to think about acceleration. The best way to understand acceleration is as a change in speed per unit of time. If a car can go from 0 to 60 mph in 6 seconds, that's an increase in speed of 60 mph in 6 seconds, or, on average, 10 mph each second — that's an acceleration of $10 \frac{\text{mph}}{\text{second}}$. It's slightly more confusing when the speed has a unit like feet/second and the unit of time for the acceleration

is also a second, because then the word *second* appears twice. But it still works like the car example. Say an object starts at rest and speeds up to 10 feet/second after 1 second, then up to 20 feet/second after 2 seconds, to 30 feet/second after 3 seconds, and so on. Its speed is increasing 10 feet/second each second, and that's an acceleration of $10 \frac{\text{feet per second}}{\text{second}}$ or $10 \frac{\text{feet/second}}{\text{second}}$. (By the way, it's helpful to write the acceleration unit in either of these ways (using a vertical fraction) as a speed over the unit of time — instead of horizontally like 10 feet per second per second or 10 feet/second/second — to emphasize that acceleration is a change in speed per unit of time.) Think of acceleration this way, not in terms of that second^2 nonsense.

Tying it all together

Note the following connections among the three graphs in Figure 12-4. The *negative* section on the graph of $A(t)$ — from $t=0$ to $t=2$ — corresponds to a *decreasing* section of the graph of $V(t)$ and a *concave down* section of the graph of $H(t)$. The *positive* interval on the graph of $A(t)$ — from $t=2$ to $t=4$ — corresponds to an *increasing* interval on the graph of $V(t)$ and a *concave up* interval on the graph of $H(t)$. When $t=2$ seconds, $A(t)$ has a *zero*, $V(t)$ has a *local minimum*, and $H(t)$ has an *inflection point*.

Related Rates — They Rate, Relatively

Say you're filling up your swimming pool and you know how fast water is coming out of your hose, and you want to calculate how fast the water level in the pool is rising. You know one rate (how fast the water is being poured in), and you want to determine another rate (how fast the water level is rising). These rates are called *related rates* because one depends on the other — the faster the water is poured in, the faster the water level will rise. In a typical related rates problem (like the one just described), the rate or rates in the given information are constant, unchanging, and you have to figure out a related rate that is changing with time. You have to determine this related rate at one particular point in time. (If this isn't crystal clear, you'll see what I mean in a minute when you work through the following problems.)

Solving these problems can be tricky at first, but with practice you'll get the hang of it. The strategies and tips are a big help — let's do three examples.

Blowing up a balloon

You're blowing up a balloon at a rate of 300 cubic inches per minute. When the balloon's radius is 3 inches, how fast is the radius increasing?

1. **Draw a diagram, labeling it with any *unchanging* measurements (there aren't any in this unusually simple problem) and making sure to assign a variable to anything in the problem that's *changing* (unless it's irrelevant to the problem). See Figure 12-6.**

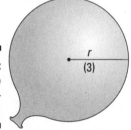

Figure 12-6:
Blowing up
a balloon —
time to party!

Notice that the radius in Figure 12-6 is labeled with the variable *r*. The radius needs a variable because as the balloon is being blown up, the radius is *changing*. I put the 3 in parentheses to emphasize that the number 3 is *not* an unchanging measurement. The problem asks you to determine something *when* the radius is 3 inches, but remember, the radius is constantly changing.

Changing or unchanging? In related rates problems, it's important to distinguish between what is changing and what is *not* changing.

The volume of the balloon is also changing, so you need a variable for volume, *V*. You could put a *V* on your diagram to indicate the changing volume, but there's really no easy way to mark part of the balloon with a *V* like you can show the radius with an *r*.

2. **List the given rate and the rate you're asked to determine as derivatives with respect to time.**

You're pumping up the balloon at 300 cubic inches per minute. That's a rate — a change in volume (cubic inches) per change in time (minutes). So,

$$\frac{dV}{dt} = 300 \text{ cubic inches per minute}$$

You have to figure out a related rate, namely, how fast the radius is changing, so

$$\frac{dr}{dt} = ?$$

3. **Write down the formula that connects the variables in the problem, V and r.**

Here's the formula for the volume of a sphere:

$$V = \frac{4}{3}\pi r^3$$

4. **Differentiate your formula with respect to time, t.**

When you differentiate in a related rates problem, all variables are treated like the ys are treated in a typical implicit differentiation problem.

$$\frac{dV}{dt} = \frac{4}{3}\pi \cdot 3r^2 \frac{dr}{dt}$$
$$= 4\pi r^2 \frac{dr}{dt}$$

You need to add the $\frac{dr}{dt}$ just like the way you add on a y' or a $\frac{dy}{dx}$ with implicit differentiation.

5. **Substitute known values for the rate and variables in the equation from Step 4, and then solve for the thing you're asked to determine.**

It's given that $\frac{dV}{dt} = 300$, and you're asked to figure out $\frac{dr}{dt}$ when $r = 3$, so plug in these numbers and solve for $\frac{dr}{dt}$.

Differentiate before you plug in. Be sure to differentiate (Step 4) *before* you plug the given information into the unknowns (Step 5).

$$\frac{dV}{dt} = 4\pi r^2 \frac{dr}{dt}$$
$$300 = 4\pi \cdot 3^2 \frac{dr}{dt}$$
$$300 = 36\pi \frac{dr}{dt}$$
$$\frac{300}{36\pi} = \frac{dr}{dt}$$
$$\frac{dr}{dt} \approx 2.65 \text{ inches per minute}$$

So the radius is increasing at a rate of about 2.65 inches per minute when the radius measures 3 inches. Think of all the balloons you've blown up since your childhood. Now you finally have the answer to the question that's been bugging you all these years.

By the way, if you plug 5 into *r* instead of 3, you get an answer of about 0.95 inches per minute. This should agree with your balloon-blowing-up experience: The bigger the balloon gets, the slower it grows. It's a good idea to check things like this every so often to see that the math agrees with common sense.

Filling up a trough

Here's another related rates problem. A trough is being filled up with swill. The trough is 10 feet long, and its cross-section is an isosceles triangle with a base of 2 feet and a height of 2 feet 6 inches (with the vertex at the bottom, of course). Swill's being poured in at a rate of 5 cubic feet per minute. When the depth of the swill is 1 foot 3 inches, how fast is the swill level rising?

1. **Draw a diagram, labeling it with any *unchanging* measurements and assigning variables to any *changing* things. See Figure 12-7.**

   Note that Figure 12-7 shows the *unchanging* dimensions of the trough, 2 feet, 2 feet 6 inches, and 10 feet, and that these dimensions do *not* have variable names like *l* for length or *h* for height. And note that the *changing* things — the height (or depth) of the swill and the width of the surface of the swill (which gets wider as the swill gets deeper) — have variable names, *h* for height and *b* for base (I say *base* instead of *width* because it's the base of the upside-down triangle shape made by the swill). The volume of the swill is also changing, so you can call that *V*.

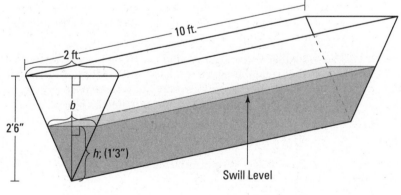

Figure 12-7:
Filling a
trough with
swill —
lunch time.

(Note: The perspective is not quite right so you can see the exact shape of the triangle.)

2. **List the given rate and the rate you're asked to figure out as derivatives with respect to *time*.**

$$\frac{dV}{dt} = 5 \text{ cubic feet per minute}$$

$$\frac{dh}{dt} = ?$$

3.a. **Write down the formula that connects the variables in the problem: *V*, *h*, and *b*.**

I'm *absolutely positive* that you remember the formula for the volume of a right prism (the shape of the swill in the trough):

$$V = (area\,of\,base)(height)$$

Note that this "base" is the base of the prism (the whole triangle at the end of the trough), not the base of the triangle which is labeled *b* in Figure 12-6. Also, this "height" is the height of the prism (the length of the trough), not the height labeled *h* in Figure 12-7. Sorry about the confusion. Deal with it.

The area of the triangular base equals $\frac{1}{2}bh$, and the "height" of the prism is 10 feet, so the formula becomes

$$V = \frac{1}{2}bh \cdot 10$$

$$V = 5bh$$

Now, unlike the formula in the balloon example, this formula contains a variable, *b*, that you don't see in your list of derivatives in Step 2. So Step 3 has a second part — getting rid of this extra variable.

3.b. **Find an equation that relates the unwanted variable, *b*, to some other variable in the problem so you can make a substitution that leaves you with only *V* and *h*.**

The triangular face of the swill in the trough is *similar* to the triangular face of the trough itself, so the base and height of these triangles are proportional. (Recall from geometry that *similar triangles* are triangles of the same shape; their sides are proportional.) Thus,

$$\frac{b}{2} = \frac{h}{2.5} \qquad \text{(Be careful: } 2'6'' \text{ is } not \text{ 2.6 feet.)}$$

$$2.5b = 2h \qquad \text{(cross multiplication)}$$

$$b = \frac{2h}{2.5}$$

$$b = 0.8h$$

Be on the lookout for similar triangles. Similar triangles come up a lot in related rates problems. Look for them whenever the problem involves a triangle, a triangular prism, or a cone shape.

Now substitute $0.8h$ for b in your formula from Step 3.a.

$$V = 5bh$$
$$V = 5 \cdot 0.8h \cdot h$$
$$V = 4h^2$$

4. **Differentiate this equation with respect to t.**

$$\frac{dV}{dt} = 8h\frac{dh}{dt} \quad \text{(the power rule with the implicit differentiation } \frac{dh}{dt}\text{)}$$

5. **Substitute known values for the rate and variable in the equation from Step 4 and then solve.**

You know that $\frac{dV}{dt} = 5$ cubic feet per minute, and you want to determine $\frac{dh}{dt}$ when h equals 1 foot 3 inches, or 1.25 feet, so plug in 5 and 1.25 and solve for $\frac{dh}{dt}$:

$$\frac{dV}{dt} = 8h\frac{dh}{dt}$$
$$5 = 8 \cdot 1.25 \cdot \frac{dh}{dt}$$
$$5 = 10 \cdot \frac{dh}{dt}$$
$$\frac{dh}{dt} = \frac{1}{2}$$

That's it. The swill's level is rising at a rate of $\frac{1}{2}$ foot per minute when the swill is 1 foot 3 inches deep. Dig in.

Fasten your seat belt: You're approaching a calculus crossroads

Ready for another common related rates problem? One car leaves an intersection traveling north at 50 mph, another is driving west toward the intersection at 40 mph. At one point, the north-bound car is three-tenths of a mile north of the intersection, and the west-bound car is four-tenths of a mile east of it. At this point, how fast is the distance between the cars changing?

1. **Do the diagram thing. See Figure 12-8.**

Variable or fixed? Before going on with this problem, I want to mention a similar problem you may run across if you're using a standard calculus textbook. It involves a ladder leaning against and sliding down a wall. Can you see that the diagram for such a ladder problem would be very similar to Figure 12-8 except that the *y*-axis would represent the wall, the *x*-axis would be the ground, and the diagonal line would be the ladder? These problems are quite similar, but there's an important difference. The distance between the cars is *changing* so the diagonal line in Figure 12-8 is labeled with a variable, *s*. A ladder, on the other hand, has a *fixed* length, so the diagonal line in your diagram for the ladder problem would be labeled with a number, not a variable.

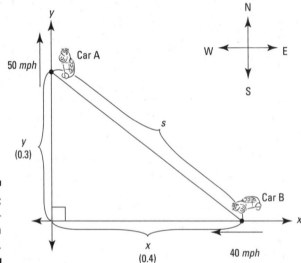

Figure 12-8:
Calculus —
it's a drive in
the country.

2. **List all given rates and the unknown rate.**

As Car A travels north, the distance *y* is growing at 50 miles per hour. That's a rate, a change in distance per change in time. So,

$$\frac{change\ in\ distance\ in\ y\ direction}{change\ in\ time} = \frac{dy}{dt} = 50 \text{ mph}$$

As Car B travels west, the distance *x* is *shrinking* at 40 miles per hour. That's a *negative* rate:

$$\frac{change\ in\ distance\ in\ x\ direction}{change\ in\ time} = \frac{dx}{dt} = -40 \text{ mph}$$

You have to figure out how fast s is changing, so,

$$\frac{\text{change in distance in } s \text{ direction}}{\text{change in time}} = \frac{ds}{dt} = ?$$

3. **Write the formula that relates the variables in the problem: x, y, and s.**

The Pythagorean theorem, $a^2 + b^2 = c^2$, will do the trick for this right triangle problem. In this problem, x and y are the legs of the right triangle, and s is the hypotenuse, so $x^2 + y^2 = s^2$.

The Pythagorean theorem is used a lot in related rates problems. If there's a right triangle in your problem, it's quite likely that $a^2 + b^2 = c^2$ is the formula you'll need.

Because this formula contains the variables x, y, and s which all appear in your list of derivatives in Step 2, you don't have to tweak this formula like you did in the trough problem.

4. **Differentiate with respect to t.**

$$s^2 = x^2 + y^2$$

$$2s\frac{ds}{dt} = 2x\frac{dx}{dt} + 2y\frac{dy}{dt} \qquad \text{(implicit differentiation with the power rule)}$$

(Remember, in a related rates problem, all variables are treated like the ys in an implicit differentiation problem.)

5. **Substitute and solve for $\dfrac{ds}{dt}$.**

$$x = 0.4, y = 0.3, \frac{dx}{dt} = -40, \frac{dy}{dt} = 50, \text{ and } s = \ \dots$$

"Holy devoid distance lacking length, Batman. How can we solve for $\dfrac{ds}{dt}$ unless we have values for the rest of the unknowns in the equation?" "Take a chill pill, Robin — just use the Pythagorean theorem again."

$$s^2 = x^2 + y^2$$
$$s^2 = 0.4^2 + 0.3^2$$
$$s^2 = 0.16 + 0.09$$
$$s^2 = 0.25$$
$$s = \pm 0.5 \qquad \text{(square rooting both sides)}$$

You can reject the negative answer because s obviously has a positive length. So $s = 0.5$.

Now plug everything into your equation:

$$2s\frac{ds}{dt} = 2x\frac{dx}{dt} + 2y\frac{dy}{dt}$$

$$2 \cdot 0.5 \cdot \frac{ds}{dt} = 2 \cdot 0.4 \cdot (-40) + 2 \cdot 0.3 \cdot 50$$

$$1 \cdot \frac{ds}{dt} = -32 + 30$$

$$\frac{ds}{dt} = -2$$

This negative answer means that the distance, s, is *decreasing*.

Thus, when car A is 3 blocks north of the intersection and car B is 4 blocks east of it, the distance between them is decreasing at a rate of 2 mph.

Chapter 13

More Differentiation Problems: Going Off on a Tangent

. . . .

In This Chapter

▶ Tangling with tangents

▶ Negotiating normals

▶ Lining up for linear approximations

▶ Profiting from business and economics problems

. . . .

*I*n this chapter, you see three more applications of differentiation: tangent and normal line problems, linear approximation problems, and economics problems. The common thread tying these problems together is the idea of a line tangent to a curve — which should come as no surprise since the meaning of the derivative of a curve is the slope of the tangent line.

Tangents and Normals: Joined at the Hip

By now you know what a line tangent to a curve looks like — if not, one or both of us has definitely dropped the ball. A *normal* line is simply a line perpendicular to a tangent line at the point of tangency. Problems involving tangents and normals are common applications of differentiation.

The tangent line problem

I bet there have been several times, just in the last month, when you've wanted to determine the location of a line through a given point that's tangent to a given curve. Here's how you do it.

Example: Determine the points of tangency of the lines through the point $(1, -1)$ that are tangent to the parabola $y = x^2$.

Solution: If you graph the parabola and plot the point, you can see that there are two ways to draw a tangent line from $(1, -1)$: up to the right and up to the left. See Figure 13-1.

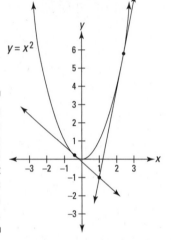

Figure 13-1:
The parabola $y = x^2$ and two tangent lines through $(1, -1)$.

The key to this problem is in the meaning of the derivative: Don't forget — *The derivative of a function at a given point is the slope of the tangent line at that point.* So, all you have to do is set the derivative of the parabola equal to the slope of the tangent line and solve:

1. **Because the equation of the parabola is $y = x^2$, you can take a general point on the parabola, (x, y), and substitute x^2 for y.**

   So, label the two points of tangency (x, x^2).

2. **Take the derivative of the parabola.**

   $$y = x^2$$
   $$y' = 2x$$

3. **Using the slope formula, $\dfrac{y_2 - y_1}{x_2 - x_1}$, set the slope of each tangent line from $(1, -1)$ to (x, x^2) equal to the derivative at (x, x^2) which is $2x$, and solve for x.**

(By the way, the math you do in this step may make more sense to you if you think of it as applying to just one of the tangent lines — say the one going up to the right — but the math actually applies to both tangent lines simultaneously.)

$$\frac{y_2 - y_1}{x_2 - x_1} \ \begin{array}{l}\text{(the slope of}\\ \text{the tangent line)}\end{array} = y' \ \text{(the derivative)}$$

$$\frac{x^2 - (-1)}{x - 1} = 2x$$
$$x^2 - (-1) = 2x(x - 1)$$
$$x^2 + 1 = 2x^2 - 2x$$
$$0 = x^2 - 2x - 1$$

$$x = \frac{2 \pm \sqrt{(-2)^2 - 4(1)(-1)}}{2 \cdot 1} \quad \text{(quadratic formula)}$$

$$= \frac{2 \pm \sqrt{8}}{2}$$

$$= \frac{2 \pm 2\sqrt{2}}{2}$$

$$= 1 \pm \sqrt{2}$$

So, the *x*-coordinates of the points of tangency are $1 + \sqrt{2}$ and $1 - \sqrt{2}$.

4. **Plug each of these *x*-coordinates into $y = x^2$ to obtain the *y*-coordinates.**

$$y = (1 + \sqrt{2})^2 \qquad\qquad\qquad y = (1 - \sqrt{2})^2$$
$$= 1 + 2\sqrt{2} + 2 \qquad\qquad\qquad = 1 - 2\sqrt{2} + 2$$
$$= 3 + 2\sqrt{2} \qquad\qquad\qquad\quad\ = 3 - 2\sqrt{2}$$

Thus, the two points of tangency are $(1 + \sqrt{2}, 3 + 2\sqrt{2})$ and $(1 - \sqrt{2}, 3 - 2\sqrt{2})$, or about $(2.4, 5.8)$ and $(-0.4, 0.2)$.

The normal line problem

Here's the companion problem to the tangent line problem in the previous section. Find the points of perpendicularity for all normal lines to the parabola, $y = \frac{1}{16}x^2$, that pass through the point $(3, 15)$.

Definition of *normal line:* A line normal to a curve at a given point is the line perpendicular to the line that's tangent to the curve at that same point.

Graph the parabola and plot the point (3, 15). Now, before you do the math, try to estimate the locations of all normal lines. How many can you see? It's fairly easy to see that, starting at (3, 15), one normal line goes down and to the right and another goes down to the left. But did you see that there's actually a second normal line that goes down to the left? No worries if you didn't see it, because when you do the math, you get all three solutions.

Making commonsense estimates enhances mathematical understanding. When doing calculus, or any math for that matter, come up with a common sense, ballpark estimate of the solution to a problem before doing the math (when possible and time permitting). This deepens your understanding of the concepts involved and provides a check to the mathematical solution. (This is a powerful math strategy — take my word for it — despite the fact that in this particular problem, most people will find at most two of the three normal lines using an eyeball estimate.)

Figure 13-2 shows the parabola and the three normal lines.

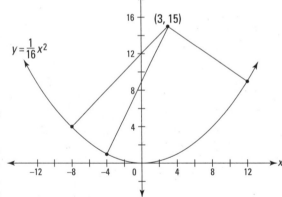

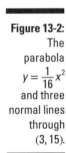

Figure 13-2: The parabola $y = \frac{1}{16}x^2$ and three normal lines through (3, 15).

Looking at the figure, you can appreciate how practical this problem is. It'll really come in handy if you happen to find yourself standing inside the curve of a parabolic wall, and you want to know the precise location of the three points on the wall where you could throw a ball and have it bounce straight back to you.

The solution is very similar to the solution of the tangent line problem, except that in this problem you use the rule for perpendicular lines:

Slopes of *perpendicular lines*. The slopes of perpendicular lines are opposite reciprocals.

Each normal line in Figure 13-2 is perpendicular to the tangent line drawn at the point where the normal meets the curve. So the slope of each normal line is the opposite reciprocal of the slope of the corresponding tangent line — which, of course, is given by the derivative. So here goes:

1. **Take a general point, (x, y), on the parabola $y = \frac{1}{16}x^2$, and substitute $\frac{1}{16}x^2$ for y.**

    So, label each point of perpendicularity $\left(x, \frac{1}{16}x^2\right)$.

2. **Take the derivative of the parabola.**

    $$y = \frac{1}{16}x^2$$
    $$y' = \frac{1}{8}x$$

3. **Using the slope formula, $\frac{y_2 - y_1}{x_2 - x_1}$, set the slope of each normal line from $(3, 15)$ to $\left(x, \frac{1}{16}x^2\right)$ equal to the opposite reciprocal of the derivative at $\left(x, \frac{1}{16}x^2\right)$, and solve for x.**

    $$\frac{\frac{1}{16}x^2 - 15}{x - 3} = -\frac{8}{x}$$

    (The derivative is $\frac{1}{8}x$ or $\frac{x}{8}$ and its opposite reciprocal is thus $-\frac{8}{x}$.)

    $$\frac{1}{16}x^3 - 15x = -8x + 24$$

    (by cross multiplication and distribution)

    $$x^3 - 112x - 384 = 0$$

    (after bringing all terms to one side and multiplying both sides by 16)

Now, there's no easy way to get *exact* solutions to this cubic (3rd degree) equation like the way the quadratic formula gives you the exact solutions to a 2nd degree equation. Instead, you can graph $y = x^3 - 112x - 384$, and the x-intercepts give you the solutions, but with this method, there's no guarantee you'll get exact solutions. (If you don't, the decimal approximations you get will be good enough.) Here, however, you luck out — actually, I had something to do with it — and get the exact solutions of $-8, -4,$ and 12. (You should graph the cubic function so you see how this works.)

4. **Plug each of these *x*-coordinates into** $y = \frac{1}{16}x^2$ **to obtain the** *y*-**coordinates.**

$$y = \frac{1}{16}(-8)^2 \qquad y = \frac{1}{16}(-4)^2 \qquad y = \frac{1}{16}(12)^2$$
$$= 4 \qquad\qquad = 1 \qquad\qquad = 9$$

Thus, the three points of normalcy are $(-8, 4), (-4, 1)$, and $(12, 9)$ — play ball!

Straight Shooting with Linear Approximations

Because ordinary functions are locally *linear* (that's straight) — and the further you zoom in on them, the straighter they look — a line tangent to a function is a good approximation of the function near the point of tangency. Figure 13-3 shows the graph of $f(x) = \sqrt{x}$ and a line tangent to the function at the point $(9, 3)$. You can see that near $(9, 3)$, the curve and the tangent line are virtually indistinguishable.

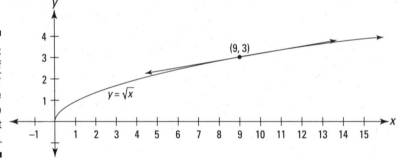

Figure 13-3:
The graph of
$f(x) = \sqrt{x}$
and a line
tangent to
the curve at
$(9, 3)$.

Determining the equation of this tangent line is a breeze. You've got a point, $(9, 3)$, and the slope is given by the derivative of *f* at 9:

$$f(x) = \sqrt{x} = x^{1/2}$$
$$f'(x) = \frac{1}{2}x^{-1/2} = \frac{1}{2\sqrt{x}} \qquad \text{(power rule)}$$
$$f'(9) = \frac{1}{2\sqrt{9}} = \frac{1}{6}$$

Now just take this slope (this derivative) of $\frac{1}{6}$ and the point $(9, 3)$, and plug them into point-slope form:

$$y - y_1 = m(x - x_1)$$
$$y - 3 = \frac{1}{6}(x - 9)$$
$$y = 3 + \frac{1}{6}(x - 9)$$

That's the equation of the line tangent to $f(x) = \sqrt{x}$ at $(9, 3)$. I suppose you may be wondering why I wrote the equation as $y = 3 + \frac{1}{6}(x - 9)$. It might seem more natural to put the 3 to the right of $\frac{1}{6}(x - 9)$, which, of course, would also be correct. And I could have simplified the equation further, writing it in $y = mx + b$ form. I explain later in this section why I wrote it the way I did.

(If you have your graphing calculator handy, graph $f(x) = \sqrt{x}$ and the tangent line. Zoom in on the point $(9, 3)$ a couple times. You'll see that — as you zoom in — the curve gets straighter and straighter and the curve and tangent line get closer and closer to each other.)

Now, say you want to approximate the square root of 10. Because 10 is pretty close to 9, and because you can see from Figure 13-3 that $f(x)$ and its tangent line are close to each other at $x = 10$, the y-coordinate of the line at $x = 10$ is a good approximation of the function value at $x = 10$, namely $\sqrt{10}$.

Just plug 10 into the line equation for your approximation:

$$y = 3 + \frac{1}{6}(x - 9)$$
$$= 3 + \frac{1}{6}(10 - 9)$$
$$= 3 + \frac{1}{6}$$
$$= 3\frac{1}{6}$$

Thus, the square root of 10 is about $3\frac{1}{6}$. This is only about 0.004 more than the exact answer of 3.1623. . . . The error is roughly a tenth of a percent.

Now I can explain why I wrote the equation for the tangent line the way I did. This form makes it easier to do the computation and easier to understand what's going on when you compute an approximation. Here's why. You know that the line goes through the point $(9, 3)$. right? And you know the slope of the line is $\frac{1}{6}$. So, you can start at $(9, 3)$ and go to the right (or left) along the line in the stair-step fashion, as shown in Figure 13-4: over 1, up $\frac{1}{6}$; over 1, up $\frac{1}{6}$; and so on. (Note that since $slope = \frac{rise}{run}$, when the run is 1 (as shown in Figure 13-4), the rise equals the slope.)

Figure 13-4:
The linear approxima-tion line and several of its points.

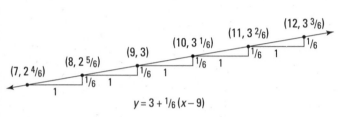

So, when you're doing an approximation, you start at a y-value of 3 and go up $\frac{1}{6}$ for each 1 you go to the right. Or if you go to the left, you go down $\frac{1}{6}$ for each 1 you go to the left. When the line equation is written in the above form, the computation of an approximation parallels this stair-step scheme.

Figure 13-4 shows the approximate values for the square roots of 7, 8, 10, 11, and 12. Here's how you come up with these values. To get to 8, for example, from $(9, 3)$, you go 1 to the left, so you go down $\frac{1}{6}$ to $2\frac{5}{6}$; or to get to 11 from $(9, 3)$, you go *two* to the right, so you go up *two*-sixths to $3\frac{2}{6}$ or $3\frac{1}{3}$. (If you go to the right *one half* to $9\frac{1}{2}$, you go up *half* of a sixth, that's a twelfth, to $3\frac{1}{12}$, the approximate square root of $9\frac{1}{2}$.)

The following list shows the size of the errors for the approximations shown in Figure 13-4. Note that the errors grow as you get further from the point of tangency $(9, 3)$. Also, the errors grow faster going down from 9 to 8 then 7, etc. than going up from 9 to 10 then 11, etc.; errors often grow faster in one direction than the other with linear approximations because of the shape of the curve.

$\sqrt{7}$: 0.8% error

$\sqrt{8}$: 0.2% error

$\sqrt{10}$: 0.1% error

$\sqrt{11}$: 0.5% error

$\sqrt{12}$: 1.0% error

Linear approximation equation: Here's the general form for the equation of the tangent line that you use for a linear approximation. The values of a function $f(x)$ can be approximated by the values of the tangent line $l(x)$ near the point of tangency, $(x_0, f(x_0))$, where

$$l(x) = f(x_0) + f'(x_0)(x - x_0)$$

This is less complicated than it looks. It's just the gussied-up calculus version of the point-slope equation of a line you've known since Algebra I, $y - y_1 = m(x - x_1)$, with the y_1 moved to the right side:

$$y = y_1 + m(x - x_1)$$

This equation and the equation for $l(x)$ differ only in the symbols used; the *meaning* of both equations — term for term — is identical. And notice how they both resemble the equation of the tangent line in Figure 13-4.

Look for algebra-calculus and geometry-calculus connections. Whenever possible, try to see the basic algebra or geometry concepts at the heart of fancy-looking calculus concepts.

Business and Economics Problems

Believe it or not, calculus is actually used in the real world of business and economics — learn calculus and increase your profits! Tell me: When you're driving around an upscale part of town and you pass by a *huge* home, what's the first thing that comes to your mind? I bet it's "Just look at that home! That guy (gal) must know calculus."

Managing marginals in economics

Look again at Figures 13-3 and 13-4 in the previous section. Recall that the derivative and thus the slope of $y = \sqrt{x}$ at $(9, 3)$ is $\frac{1}{6}$, and that the tangent line at this point can be used to approximate the function near the point of tangency. So, as you go over 1 from 9 to 10 along the function itself, you go up *about* $\frac{1}{6}$. And, thus, $\sqrt{10}$ is about $\frac{1}{6}$ more than $\sqrt{9}$. The mathematics of *marginals* works exactly the same way.

Marginal cost, marginal revenue, and marginal profit work a lot like linear approximation. *Marginal cost, marginal revenue,* and *marginal profit* all involve how much a function goes up (or down) as you go over 1 to the right — just like with linear approximation.

Say you've got a cost function that gives you the total cost, $C(x)$ of producing x items. See Figure 13-5.

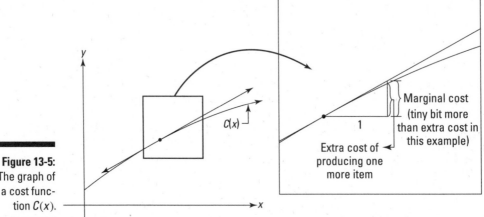

Figure 13-5:
The graph of
a cost func-
tion $C(x)$.

Look at the blown-up square on the right in the figure. The derivative of $C(x)$ at the point of tangency gives you the slope of the tangent line and thus the amount you go up as you go 1 to the right along the tangent line. (This amount is labeled in the figure as *marginal cost.*) Going 1 to the right along the cost function itself shows you the increase in cost of producing one more item. (This is labeled as the *extra cost.*) Because the tangent line is a good approximation of the cost function, the derivative of C — called the *marginal cost* — is the *approximate* increase in cost of producing one more item. Marginal revenue and marginal profit work the same way.

Definition of *marginal cost, marginal revenue,* and *marginal profit*:

Marginal cost equals the derivative of the cost function.

Marginal revenue equals the derivative of the revenue function.

Marginal profit equals the derivative of the profit function.

Before doing an example involving marginals, there's one more piece of business to take care of. A *demand function* tells you how many items will be purchased (what the demand will be) given the price. The lower the price, of course, the higher the demand; and the higher the price, the lower the demand. You'd think that the number purchased should be a function of the price — input a price and find out how many items people will buy at that price — but traditionally, a demand function is written the other way around. The price is expressed as a function of the number demanded. I know that seems a bit odd, but don't sweat it — the function works either way. Think of it like this: If a retailer wants to sell a given number of items, the demand function tells the retailer what he or she should set the selling price at.

Okay, so here's the example. A widget manufacturer determines that the demand function for his widgets is

$$p = \frac{1000}{\sqrt{x}}$$

where p is the price of a widget and x is the number of widgets demanded. The cost of producing x widgets is given by the following cost function:

$$C(x) = 10x + 100\sqrt{x} + 10{,}000$$

Determine the marginal cost, marginal revenue, and marginal profit at $x = 100$ widgets. Also, how many widgets should be manufactured, and what should they be sold for to produce the maximum profit, and what is that maximum profit? (If you get through all this, I'll nominate you for the Nobel Prize in Economics.)

Marginal cost

Marginal cost is the derivative of the cost function, so take the derivative and evaluate it at $x = 100$:

$$C(x) = 10x + 100\sqrt{x} + 10{,}000$$

$$C'(x) = 10x + \frac{50}{\sqrt{x}} \quad \text{(power rule)}$$

$$C'(100) = 10 + \frac{50}{\sqrt{100}}$$

$$= 10 + \frac{50}{10}$$

$$= 15$$

Thus, the marginal cost at $x = 100$ is $15 — this is the approximate cost of producing the 101st widget.

Marginal revenue

Revenue, $R(x)$, equals the number of items sold, x, times the price, p:

$$R(x) = x \cdot p$$

$$= x \cdot \frac{1000}{\sqrt{x}} \qquad \text{(using the above demand function)}$$

$$= \frac{1000x}{\sqrt{x}} \cdot \frac{\sqrt{x}}{\sqrt{x}} \qquad \text{(rationalizing the denominator)}$$

$$= \frac{1000x\sqrt{x}}{x}$$

$$= 1000\sqrt{x}$$

Marginal revenue is the derivative of the revenue function, so take the derivative of $R(x)$ and evaluate it at $x = 100$:

$$R(x) = 1000\sqrt{x}$$

$$R'(x) = \frac{500}{\sqrt{x}} \qquad \text{(power rule)}$$

$$R'(100) = \frac{500}{\sqrt{100}}$$

$$= 50$$

Thus, the approximate revenue from selling the 101st widget is \$50.

Marginal profit

Profit, $P(x)$, equals revenue minus cost. So,

$$P(x) = R(x) - C(x)$$

$$= 1000\sqrt{x} - (10x + 100\sqrt{x} + 10{,}000)$$

$$= -10x + 900\sqrt{x} - 10{,}000$$

Marginal profit is the derivative of the profit function, so take the derivative of $P(x)$, and evaluate it at $x = 100$:

$$P(x) = -10x + 900\sqrt{x} - 10{,}000$$

$$P'(x) = -10 + \frac{450}{\sqrt{x}} \qquad \text{(power rule)}$$

$$P'(100) = -10 + \frac{450}{\sqrt{100}}$$

$$= -10 + 45$$

$$= 35$$

Selling the 101st widget brings in an approximate profit of $35.

Marginal profit short cuts: Did you notice either of the two shortcuts you could have taken here? First, you can use the fact that

$$P'(x) = R'(x) - C'(x)$$

to determine $P'(x)$ directly, without first determining $P(x)$. Then, after getting $P'(x)$, you just plug 100 into x for your answer.

And, if all you want to know is $P'(100)$, you can use the following really short shortcut:

$$P'(100) = R'(100) - C'(100)$$
$$= 50 - 15$$
$$= 35$$

This is common sense. If it costs you about $15 to produce the 101st widget and you sell it for about $50, then your profit is about $35.

I did it the long way because you need both the profit function, $P(x)$, and the marginal profit function, $P'(x)$, for the problems below.

Maximum profit

To determine maximum profit, set the derivative of profit — that's marginal profit — equal to zero, solve for x, and then plug the result into the profit function:

$$P'(x) = -10 + \frac{450}{\sqrt{x}}$$
$$0 = -10 + \frac{450}{\sqrt{x}}$$
$$10 = \frac{450}{\sqrt{x}}$$
$$10\sqrt{x} = 450$$
$$\sqrt{x} = 45$$
$$x = 2025$$

So, maximum profit occurs when 2,025 widgets are sold. Plug this into $P(x)$:

$$P(x) = -10x + 900\sqrt{x} - 10{,}000$$
$$P(2025) = -10 \cdot 2025 + 900\sqrt{2025} - 10{,}000$$
$$= -20{,}250 + 900 \cdot 45 - 10{,}000$$
$$= 10{,}250$$

Thus, the maximum profit is $10,250. (Extra credit: Did you see where I got a bit lazy here? The derivative of the profit function is zero at $x = 2025$, but that doesn't guarantee that there's a max at that x-value. There could instead be a min or an inflection point there. You could use either the first or second derivative test (see Chapter 11) to show that it's actually a max. But I just peaked at a graph of the profit function and saw that it's sort of an upside-down cup shape, so I knew that there was a max at the top of the cup at $x = 2025$.)

Finally, plug the number sold into the demand function to determine the profit-maximizing price:

$$p = \frac{1000}{\sqrt{x}}$$

$$p = \frac{1000}{\sqrt{2025}}$$

$$\approx 22.22$$

So, the maximum profit of $10,250 occurs when the price is set at $22.22. At this price, 2,025 widgets will be sold. Figure 13-6 sums up these results. Note that because profit equals revenue minus cost, the vertical distance or gap between the revenue and cost functions at a given x-value gives the profit at that x-value. Maximum profit occurs where the gap is greatest.

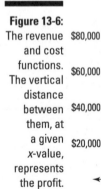

Figure 13-6:
The revenue and cost functions. The vertical distance between them, at a given x-value, represents the profit.

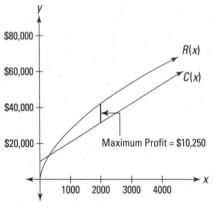

(Note that although the scale of this graph makes $C(x) = 10x + 100\sqrt{x} + 10{,}000$ look like a straight line, its middle term of $100\sqrt{x}$ means that it is not exactly straight.)

And here's another thing. Because maximum profit occurs where $P'(x) = 0$, and because $P'(x) = R'(x) - C'(x)$, it follows that the profit will be greatest where $0 = R'(x) - C'(x)$ — in other words, where $R'(x) = C'(x)$. And when $R'(x) = C'(x)$, the slopes of the functions' tangent lines are equal. So, if you were to draw tangent lines to $R(x)$ and $C(x)$ where the gap between the two is greatest, these tangents would be parallel. Right about now you're probably thinking something like — *Such symmetry, such simple elegance, such beauty! Verily, the mathematics muse seduces the heart as much as the mind.* Yeah, it's nice all right, but let's not get carried away.

Part V
Integration and Infinite Series

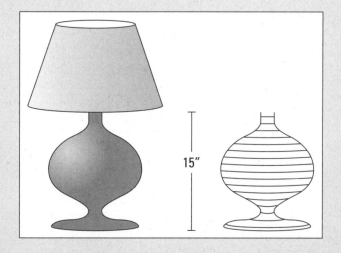

15"

In this part . . .

✔ The meaning of integration: Integration is basically just fancy calculus addition. It works by sort of slicing up something into tiny bits (actually infinitesimal bits) and then adding up the bits to get the total. This allows you to find the total of things — say, the total volume of some weird bell-shaped object — that can't be handled by simple, pre-calculus formulas.

✔ The fundamental theorem of calculus: Integration is basically differentiation in reverse — namely, *antidifferentiation*. There's an intimate yin/yang connection between integration and differentiation which this part looks at from several angles.

✔ Techniques for calculating antiderivatives: Substitution, integration by parts, trig integrals, and partial fractions.

✔ Integration word problems: Calculating the area between curves, finding volume with the washer method, computing arc length and surface area, and improper integrals.

✔ L'Hôpital's rule: A nice trick for solving limit problems.

✔ Taming infinity: Ten tests for the convergence or divergence of *infinite series*.

Chapter 14

Intro to Integration and Approximating Area

. . . .

In This Chapter

▶ Integrating — adding it all up

▶ Approximating areas and sizing up sigma sums

▶ Using the definite integral to get exact areas

▶ Totaling up trapezoids

▶ Simpson's rule: Calculus for Bart and Homer

. . . .

Since you're still reading this book, I presume that means you survived differentiation (Chapters 9 through 13). Now you begin the second major topic in calculus: integration. Just as two simple ideas lie at the heart of differentiation — *rate* (like *miles per hour*) and the steepness or *slope* of a curve — integration can also be understood in terms of two simple ideas — *adding up* small pieces of something and the *area* under a curve. In this chapter, I introduce you to these two fundamental concepts.

Integration: Just Fancy Addition

Consider the lamp on the left in Figure 14-1. Say you want to determine the volume of the lamp's base. Why would you want to do that? Beats me. Anyway, a formula for the volume of such a weird shape doesn't exist, so you can't calculate the volume directly.

You can, however, calculate the volume with integration. Imagine that the base is cut up into thin, horizontal slices as shown on the right in Figure 14-1.

Figure 14-1:
A lamp with
a curvy
base and
the base
cut into thin
horizontal
slices.

15″

Can you see that each slice is shaped like a thin pancake? Now, because there *is* a formula for the volume of a pancake (a pancake is just a very short cylinder), you can determine the total volume of the base by simply calculating the volume of each pancake-shaped slice and then adding up the volumes. That's integration in a nutshell.

But, of course, if that's all there was to integration, there wouldn't be such a fuss about it — certainly not enough to vault Newton, Leibnitz, and the other all-stars into the mathematics hall of fame. What makes integration one of the great achievements in the history of mathematics is that — to continue with the lamp example — it gives you the *exact* volume of the lamp's base by sort of cutting it into an *infinite* number of *infinitely* thin slices. Now *that* is something. If you cut the lamp into fewer than an infinite number of slices, you can get only a very good approximation of the total volume — not the exact answer — because each pancake-shaped slice would have a weird, curved edge which would cause a small error when computing the volume of the slice with the cylinder formula.

Integration has an elegant symbol: $\int$. You've probably seen it before — maybe in one of those cartoons with some Einstein guy in front of a blackboard filled with indecipherable gobbledygook. Soon, this will be *you*. That's right: You'll be filling up pages in your notebook with equations containing the integration symbol. Onlookers will be amazed and envious.

You can think of the integration symbol as just an elongated *S* for "sum up." So, for our lamp problem, you can write

$$\int_{bottom}^{top} dB = B$$

where *dB* means a little bit of the base — actually an infinitely small piece. So the equation just means that if you sum up all the little pieces of the base from the *bottom* to the *top,* the result is B, the volume of the whole base.

This is a bit oversimplified — I can hear the siren of the math police now — but it's a good way to think about integration. By the way, thinking of dB as a little or infinitesimal piece of B is an idea you saw before with differentiation (see Chapter 9), where the derivative or slope, $\dfrac{dy}{dx}$, is equal to the ratio of a little bit of y to a little bit of x, as you shrink the $\dfrac{rise}{run}$ stair step down to an infinitesimal size (see Figure 9-13). Thus, both differentiation and integration involve infinitesimals.

So, whenever you see something like

$$\int_{a}^{b} little\ piece\ of\ mumbo\ jumbo$$

it just means that you add up all the little (infinitesimal) pieces of the mumbo jumbo from a to b to get the total of all of the mumbo jumbo from a to b. Or you might see something like

$$\int_{t=0\,sec.}^{t=20\,sec.} little\ piece\ of\ distance$$

which means to add up the little pieces of distance traveled between 0 and 20 seconds to get the total distance traveled during that time span.

To sum up — that's a pun! — the mathematical expression to the right of the integration symbol stands for a little bit of something, and integrating such an expression means to add up all the little pieces between some starting point and some ending point to determine the total between the two points.

Finding the Area Under a Curve

As I discuss in Chapter 9, the most fundamental meaning of a derivative is that it's a rate, a *this per that* like *miles per hour,* and that when you graph the *this* as a function of the *that* (like *miles* as a function of *hours)*, the derivative becomes the slope of the function. In other words, the derivative is a rate, which on a graph appears as a slope.

It works in a similar way with integration. The most fundamental meaning of integration is to add up (you might be adding up distances or volumes, for example). And when you depict integration on a graph, you can see the adding up process as a summing up of little bits of area to arrive at the total area under a curve. Consider Figure 14-2.

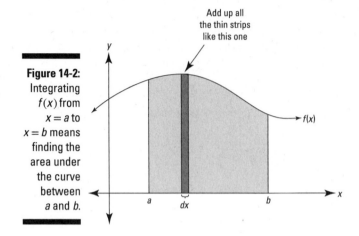

Figure 14-2: Integrating $f(x)$ from $x = a$ to $x = b$ means finding the area under the curve between a and b.

Add up all the thin strips like this one

The shaded area in Figure 14-2 can be calculated with the following integral:

$$\int_a^b f(x)dx$$

Look at the thin rectangle in Figure 14-2. It has a height of $f(x)$ and a width of dx (a little bit of x), so its area (*length* times *width*, of course) is given by $f(x) \cdot dx$. The above integral tells you to add up the areas of all the narrow rectangular strips between a and b under the curve $f(x)$. As the strips get narrower and narrower, you get a better and better estimate of the area. The power of integration lies in the fact that it gives you the *exact* area by sort of adding up an infinite number of infinitely thin rectangles.

If you're doing a problem where both the x and y axes are labeled in a unit of length, say, *feet*, then each thin rectangle measures so many feet by so many feet, and its area — *length* times *width* — is some number of *square feet*. In this case, when you integrate to get the total area under the curve between a and b, your final answer will be an amount of — what else? — area. But you can use this adding-up-areas-of-rectangles scheme to add up tiny bits of anything — distance, volume, or energy, for example. In other words, the area under the curve doesn't have to stand for an actual area.

If, for example, the units on the x-axis are *hours* and the y-axis is labeled in *miles per hour,* then, because *rate* times *time* equals *distance* (and because $\frac{miles}{hour} \cdot hours = miles$), the area of each rectangle represents an amount of distance (in miles), and the total area gives you the total distance traveled during the given time interval. Or if the x-axis is labeled in *hours* and the y-axis in *kilowatts* of electrical power — in which case the curve gives power usage as a function of time — then the area of each rectangular strip (*kilowatts* times *hours*) represents a number of *kilowatt-hours* of energy. In that case, the total area under the curve gives you the total number of kilowatt-hours of energy consumption between two points in time.

Figure 14-3 shows how you would do the lamp volume problem — from earlier in this chapter — by adding up areas. In this graph, the function $A(h)$ gives the cross-sectional *area* of a thin pancake slice of the lamp as a function of its height measured from the bottom of the lamp. So this time, the *h*-axis is labeled in *inches* (that's *h* as in *height* from the bottom of the lamp), and the *y*-axis is labeled in *square inches,* and thus each thin rectangle has a width measured in inches and a height measured in square inches. Its area, therefore, represents *inches* times *square inches*, or *cubic inches* of volume.

Figure 14-3: This shaded *area* gives you the *volume* of the base of the lamp in Figure 14-1.

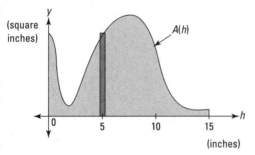

The area of the thin rectangle in Figure 14-3 represents the *volume* of the thin pancake slice of the lamp 5 inches up from the bottom of the base. The total shaded area and thus the volume of the lamp's base is given by the following integral:

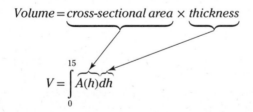

$$Volume = cross\text{-}sectional\ area \times thickness$$

$$V = \int_{0}^{15} A(h)dh$$

This integral tells you to add up the volumes of all the thin pancake slices from 0 to 15 inches (that is, from the bottom to the top of the lamp's base), each slice having a volume given by $A(h)$ (its cross-sectional area) times *dh* (its height or thickness). (By the way, Figure 14-3 resembles the left half of the lamp's base (tilted on its side), but it's not that. It has a similar shape because where the lamp base is wide, the corresponding circular slice has a large area.)

Okay, enough of this introductory stuff. In the next section, you actually calculate some areas.

Approximating Area

Before explaining how to calculate exact areas, I want to show you how to approximate areas. The approximation method is useful not only because it lays the groundwork for the exact method — integration — but because for some curves, integration is impossible, and an approximation of area is the best you can do.

Approximating area with left sums

Say you want the exact area under the curve $f(x) = x^2 + 1$ between $x = 0$ and $x = 3$. See the shaded area on the graph on the left in Figure 14-4.

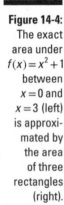

Figure 14-4: The exact area under $f(x) = x^2 + 1$ between $x = 0$ and $x = 3$ (left) is approximated by the area of three rectangles (right).

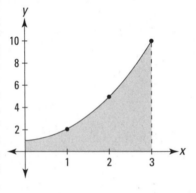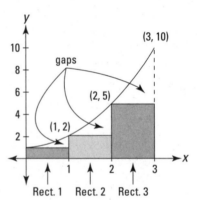

You can get a rough estimate of the total area by drawing three rectangles under the curve, as shown on the right in Figure 14-4, and then adding up their areas.

The rectangles in Figure 14-4 represent a so-called *left sum* because the height of each rectangle is determined by where the upper *left* corner of each rectangle touches the curve. Each rectangle has a width of 1 and the height of each is given by the height of the function at the rectangle's left edge. So, rectangle number 1 has a height of $f(0) = 0^2 + 1 = 1$; its area (*length · width* or *height · width*) is thus $1 \cdot 1$, or 1. Rectangle 2 has a height of $f(1) = 1^2 + 1 = 2$, so its area is $2 \cdot 1$, or 2. And rectangle 3 has a height of $f(2) = 2^2 + 1 = 5$, so its area is $5 \cdot 1$, or 5. Adding these three areas gives you a total of $1 + 2 + 5$, or 8. You can see that this is an underestimate of the total area under the curve because of the three gaps between the rectangles and the curve shown in Figure 14-4.

For a better estimate, double the number of rectangles to six. Figure 14-5 shows six "left" rectangles under the curve and also how the six rectangles begin to fill up the three gaps you see in Figure 14-4.

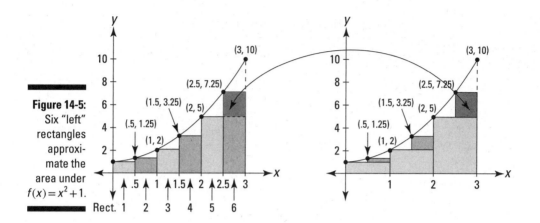

Figure 14-5: Six "left" rectangles approximate the area under $f(x) = x^2 + 1$.

See the three small shaded rectangles in the graph on the right in Figure 14-5? They sit on top of the three rectangles from Figure 14-4 and represent how much the area estimate has improved by using six rectangles instead of three.

Now total up the areas of the six rectangles. Each has a width of 0.5 and the heights are $f(0), f(0.5), f(1), f(1.5)$, and so on. I'll spare you the arithmetic. Here's the total: $0.5 + 0.625 + 1 + 1.625 + 2.5 + 3.625 = 9.875$. This is a better estimate, but it's still an underestimate because of the six small gaps you can see on the left graph in Figure 14-5.

Table 14-1 shows the area estimates given by 3, 6, 12, 24, 48, 96, 192, and 384 rectangles. You don't have to double the number of rectangles each time like I've done here. You can use any number of rectangles of equal width that you want. I just like the doubling scheme because, with each doubling, the gaps are plugged up more and more in the way shown in Figure 14-5. Any guesses as to where the estimates in Table 14-1 are headed? Looks like 12 to me.

Table 14-1	Estimates of the Area Under $f(x) = x^2 + 1$ Given by Increasing Numbers of "Left" Rectangles
Number of Rectangles	**Area Estimate**
3	8
6	9.875
12	~10.906
24	~11.445
48	~11.721
96	~11.860
192	~11.930
384	~11.965

Here's the fancy-pants formula for a left-rectangle sum:

The left rectangle rule: You can approximate the exact area under a curve between a and b, $\int_a^b f(x)dx$, with a sum of *left* rectangles of equal width given by the following formula. In general, the more rectangles, the better the estimate.

$$L_n = \frac{b-a}{n}\left[f(x_0)+f(x_1)+f(x_2)+\ldots\ldots+f(x_{n-1})\right],$$

where n is the number of rectangles, $\frac{b-a}{n}$ is the width of each rectangle, x_0 through x_{n-1} are the x-coordinates of the left edges of the n rectangles, and the function values are the heights of the rectangles.

I'd better explain this formula a bit. Look back to the six rectangles shown in Figure 14-5. The width of each rectangle equals the length of the total span from 0 to 3 (which of course is $3-0$, or 3) divided by the number of rectangles, 6. That's what the $\frac{b-a}{n}$ does in the formula.

Now, what about those xs with the subscripts? The x-coordinate of the *left* edge of rectangle 1 in Figure 14-5 is called x_0, the *right* edge of rectangle 1 (which is the same as the left edge of rectangle 2) is at x_1, the right edge of rectangle 2 is at x_2, the right edge of rectangle 3 is at x_3, and so on all the way up to the right edge of rectangle 6, which is at x_6. For the six rectangles in Figure 14-5, x_0 is 0, x_1 is 0.5, x_2 is 1, x_3 is 1.5, x_4 is 2, x_5 is 2.5, and x_6 is 3. The heights of the six left rectangles in Figure 14-5 occur at their left edges, which are at x_0 through x_5. You don't use the right edge of the last rectangle, x_6, in a left sum. That's why the list of function values in the formula stops at x_{n-1}. This all becomes clearer — cross your fingers — when you look at the formula for *right* rectangles in the next section.

Here's how to use the formula for the six rectangles in Figure 14-5:

$$L_6 = \frac{3-0}{6}\left[f(x_0)+f(x_1)+f(x_2)+f(x_3)+f(x_4)+f(x_5)\right]$$

$$= \frac{1}{2}\left[f(0)+f(0.5)+f(1)+f(1.5)+f(2)+f(2.5)\right]$$

$$= \frac{1}{2}\left(1+1.25+2+3.25+5+7.25\right)$$

$$= \frac{1}{2}\left(19.75\right)$$

$$= 9.875$$

Note that had I distributed the width of $\frac{1}{2}$ to each of the heights after the third line in the solution, you'd have seen the sum of the areas of the six rectangles — which you saw two paragraphs below Figure 14-5. The formula just uses the shortcut of first adding up the heights and then multiplying by the width.

Approximating area with right sums

Now estimate the same area under $f(x) = x^2 + 1$ from 0 to 3 with *right* rectangles. This method works like the left sum method except each rectangle is drawn so that its *right* upper corner touches the curve. See Figure 14-6.

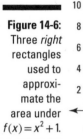

Figure 14-6: Three *right* rectangles used to approximate the area under $f(x) = x^2 + 1$.

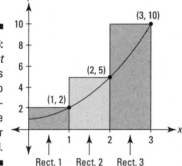

The heights of the three rectangles in Figure 14-6 are given by the function values at their *right* edges: $f(1) = 2$, $f(2) = 5$, and $f(3) = 10$. Each rectangle has a width of 1, so the areas are 2, 5, and 10, which total 17. You don't have to be a rocket scientist to see that this time you get an *over*estimate of the actual area under the curve, as opposed to the *under*estimate that you get with the left-rectangle method I detail in the previous section (more on that in a minute). Table 14-2 shows the improving estimates you get with more and more right rectangles.

Table 14-2 **Estimates of the Area Under $f(x) = x^2 + 1$ Given by Increasing Numbers of "Right" Rectangles**

Number of Rectangles	Area Estimate
3	17
6	14.375
12	~13.156
24	~12.570
48	~12.283
96	~12.141
192	~12.070
384	~12.035

Looks like these estimates are also headed toward 12. Here's the formula for a right rectangle sum.

The right rectangle rule: You can approximate the exact area under a curve between a and b, $\int_{a}^{b} f(x)dx$, with a sum of *right* rectangles given by the following formula. In general, the more rectangles, the better the estimate.

$$R_n = \frac{b-a}{n} \left[f(x_1) + f(x_2) + f(x_3) + \ldots\ldots + f(x_n) \right],$$

where n is the number of rectangles, $\frac{b-a}{n}$ is the width of each rectangle, x_1 through x_n are the x-coordinates of the right edges of the n rectangles, and the function values are the heights of the rectangles.

If you compare this formula to the one for a left rectangle sum, you get the complete picture about those subscripts. The two formulas are the same except for one thing. Look at the sums of the function values in both formulas. The right sum formula has one value, $f(x_n)$, that the left sum formula doesn't have, and the left sum formula has one value, $f(x_0)$, that the right sum formula doesn't have. All the function values between those two appear in both formulas. You can get a better handle on this by comparing the three left rectangles from Figure 14-4 to the three right rectangles from Figure 14-6. Their areas and totals, which we earlier calculated, are

Three left rectangles: $1+2+5=8$
Three right rectangles: $2+5+10=17$

The values used in the sums of the areas are the same except for the left-most left rectangle value and the right-most right rectangle value. Both sums include rectangles with areas 2 and 5. If you look at how the rectangles are constructed, you can see that the second and third rectangles in Figure 14-4 are the same as the first and second rectangles in Figure 14-6.

Approximating area with midpoint sums

A third way to approximate areas with rectangles is to make each rectangle cross the curve at the midpoint of its top side. A midpoint sum is usually a *much* better estimate of area than either a left or a right sum. Figure 14-7 shows why.

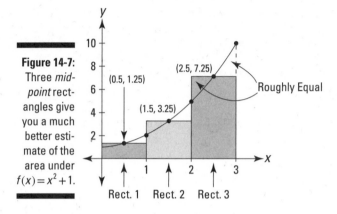

Figure 14-7:
Three *mid-point* rect-angles give
you a much
better esti-
mate of the
area under
$f(x) = x^2 + 1$.

You can see in Figure 14-7 that the part of each rectangle that's above the curve
looks about the same size as the gap between the rectangle and the curve. A
midpoint sum produces such a good estimate because these two errors roughly
cancel out each other.

For the three rectangles in Figure 14-7, the widths are 1 and the heights are
$f(0.5) = 1.25$, $f(1.5) = 3.25$, and $f(2.5) = 7.25$. The total area comes to 11.75.
Table 14-3 lists the midpoint sums for the same number of rectangles used in
Tables 14-1 and 14-2.

**Table 14-3   Estimates of the Area Under $f(x) = x^2 + 1$ Given by
Increasing Numbers of "Midpoint" Rectangles**

Number of Rectangles	Area Estimate
3	11.75
6	11.9375
12	~11.9844
24	~11.9961
48	~11.9990
96	~11.9998
192	~11.9999
384	~11.99998

If you had any doubts that the left and right sums in Tables 14-1 and 14-2
were heading to 12, Table 14-3 should dispel them. Spoiler alert: Yes, in fact,
the exact area is 12. (I show you how to calculate that in about 5 or 6 pages.)
And to see how much faster the midpoint approximations approach the exact

answer of 12 than the left or right approximations, compare the three tables. The error with 6 midpoint rectangles is about the same as the error with 192 left or right rectangles! Here's the mumbo jumbo.

The midpoint rule: You can approximate the exact area under a curve between a and b, $\int_a^b f(x)dx$, with a sum of *midpoint* rectangles given by the following formula. In general, the more rectangles, the better the estimate.

$$M_n = \frac{b-a}{n}\left[f\left(\frac{x_0+x_1}{2}\right)+f\left(\frac{x_1+x_2}{2}\right)+f\left(\frac{x_2+x_3}{2}\right)+\ldots\ldots+f\left(\frac{x_{n-1}+x_n}{2}\right)\right],$$

where n is the number of rectangles, $\frac{b-a}{n}$ is the width of each rectangle, x_0 through x_n are the $n+1$ evenly spaced points from a to b, and the function values are the heights of the rectangles.

Definition of *Riemann sum:* All three sums — left, right, and midpoint — are called Riemann sums, after the great German mathematician Bernhard Riemann (1826-66). Basically, any approximating sum made up of rectangles is a Riemann sum, including weird sums consisting of rectangles of unequal width. Luckily, you won't have to deal with those in this book or your calculus course.

The left, right, and midpoint sums in Tables 14-1, 14-2, and 14-3 are all heading toward 12, and if you could slice up the area into an infinite number of rectangles, you'd get the exact area of 12. But I'm getting ahead of myself.

Getting Fancy with Summation Notation

Before I get to the formal definition of the *definite integral* — that's the incredible calculus tool that sort of cuts up an area into an infinite number of rectangles and thereby gives you the *exact* area — there's one more thing to take care of: summation notation.

Summing up the basics

For adding up long series of numbers like the rectangle areas in a left, right, or midpoint sum, summation or *sigma* notation comes in handy. Here's how it works. Say you wanted to add up the first 100 multiples of 5 — that's from 5 to 500. You could write out the sum like this:

$$5+10+15+20+25+\ldots\ldots+490+495+500$$

But with sigma notation (sigma, $\sum$, is the 18th letter of the Greek alphabet) the sum is much more condensed and efficient, and, let's be honest, it looks pretty cool:

$$\sum_{i=1}^{100} 5i$$

This notation just tells you to plug 1 in for the i in $5i$, then plug 2 into the i in $5i$, then 3, then 4, and so on, up to 100. Then you add up the results. So that's $5 \cdot 1$ plus $5 \cdot 2$ plus $5 \cdot 3$, and so on, up to $5 \cdot 100$. This produces the same thing as writing out the sum the long way. By the way, the letter i has no significance. You can write the sum with a j, $\sum_{j=1}^{100} 5j$, or any other letter.

Here's one more. If you want to add up $10^2 + 11^2 + 12^2 + \ldots\ldots + 29^2 + 30^2$, you can write the sum with sigma notation as follows:

$$\sum_{k=10}^{30} k^2$$

There's really nothing to it.

Writing Riemann sums with sigma notation

You can use sigma notation to write out the right-rectangle sum for the curve $x^2 + 1$ we've been looking at. By the way, you don't need sigma notation for the math that follows. It's just a "convenience" — oh, sure. Cross your fingers and hope your teacher decides not to cover this. It gets pretty gnarly.

Recall the formula for a right sum from the earlier "Approximating area with right sums" section:

$$R_n = \frac{b-a}{n}\left[f(x_1) + f(x_2) + f(x_3) + \ldots\ldots + f(x_n)\right]$$

Here's the same formula written with sigma notation:

$$R_n = \sum_{i=1}^{n}\left[f(x_i) \cdot \frac{b-a}{n}\right]$$

(Note that I could have written this instead as $R_n = \dfrac{b-a}{n} \sum\limits_{i=1}^{n} f(x_i)$, which would have more nicely mirrored the above formula where the $\dfrac{b-a}{n}$ is on the outside. Either way is fine — they're equivalent — but I chose to keep the $\dfrac{b-a}{n}$ on the inside so that the $\sum$ sum is actually a sum of rectangles. In other words, with the $\dfrac{b-a}{n}$ on the inside, the expression after the $\sum$ symbol, $f(x_i) \cdot \dfrac{b-a}{n}$, which the $\sum$ symbol tells you to add up, is the area of each rectangle, namely *height* times *base*.)

Now work this out for the six right rectangles in Figure 14-8.

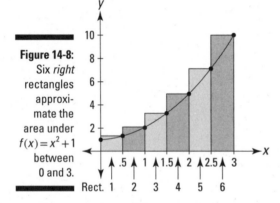

You're figuring the area under $x^2 + 1$ between $x = 0$ and $x = 3$ with six rectangles, so the width of each, $\dfrac{b-a}{n}$, is $\dfrac{3-0}{6}$, or $\dfrac{3}{6}$, or $\dfrac{1}{2}$. So now you've got

$$R_6 = \sum_{i=1}^{6} \left[f(x_i) \cdot \dfrac{1}{2} \right]$$

Next, because the width of each rectangle is $\dfrac{1}{2}$, the right edges of the six rectangles fall on the first six multiples of $\dfrac{1}{2}$: 0.5, 1, 1.5, 2, 2.5, and 3. These numbers are the x-coordinates of the six points x_1 through x_6; they can be generated by the expression $\dfrac{1}{2}i$, where i equals 1 through 6. You can check that this works by plugging 1 in for i in $\dfrac{1}{2}i$, then 2, then 3, up to 6. So now you can replace the x_i in the formula with $\dfrac{1}{2}i$, giving you

$$R_6 = \sum_{i=1}^{6} \left[f\left(\frac{1}{2}i\right) \cdot \frac{1}{2} \right]$$

Our function, $f(x)$, is $x^2 + 1$ so $f\left(\frac{1}{2}i\right) = \left(\frac{1}{2}i\right)^2 + 1$, and so now you can write

$$R_6 = \sum_{i=1}^{6} \left[\left(\left(\frac{1}{2}i\right)^2 + 1 \right) \cdot \frac{1}{2} \right]$$

If you plug 1 into i, then 2, then 3, and so on up to 6 and do the math, you get the sum of the areas of the rectangles in Figure 14-8. This sigma notation is just a fancy way of writing the sum of the six rectangles.

Are we having fun? Hold on, it gets worse — sorry. Now you're going to write out the general sum for an unknown number, n, of right rectangles. The total span of the area in question is 3, right? You divide this span by the number of rectangles to get the width of each rectangle. With 6 rectangles, the width of each is $\frac{3}{6}$; with n rectangles, the width of each is $\frac{3}{n}$. And the right edges of the n rectangles are generated by $\frac{3}{n}i$, for i equals 1 through n. That gives you

$$R_n = \sum_{i=1}^{n} \left[f\left(\frac{3}{n}i\right) \cdot \frac{3}{n} \right]$$

Or, because $f(x) = x^2 + 1$,

$$R_n = \sum_{i=1}^{n} \left[\left(\left(\frac{3}{n}i\right)^2 + 1 \right) \cdot \frac{3}{n} \right]$$

$$= \sum_{i=1}^{n} \left[\left(\frac{9i^2}{n^2} + 1 \right) \cdot \frac{3}{n} \right]$$

$$= \sum_{i=1}^{n} \left[\frac{27i^2}{n^3} + \frac{3}{n} \right]$$

$$= \sum_{i=1}^{n} \frac{27i^2}{n^3} + \sum_{i=1}^{n} \frac{3}{n} \qquad \text{(Take my word for it.)}$$

$$= \frac{27}{n^3} \sum_{i=1}^{n} i^2 + \frac{3}{n} \sum_{i=1}^{n} 1$$

For this last step, you pull the $\frac{27}{n^3}$ and the $\frac{3}{n}$ through the summation symbols — you're allowed to pull out anything except for a function of i, the so-called *index of summation*. Also, the second summation in the last step has just a 1 after it and no i. So there's nowhere to plug in the values of i. This situation may seem a bit weird, but all you do is add up n 1s, which equals n (I do this below).

You've now arrived at a critical step. With a sleight of hand, you're going to turn the above Riemann sum into a formula in terms of n. (This formula is what you use in the next section to obtain the exact area under the curve.)

Now, as almost no one knows, the sum of the first n square numbers, $1^2 + 2^2 + 3^2 + \ldots + n^2$, equals $\frac{n(n+1)(2n+1)}{6}$. (By the way, this 6 has nothing to do with the fact that we used 6 rectangles a couple pages back.) So, you can substitute that expression for the $\sum\limits_{i=1}^{n} i^2$ in the last line of the sigma notation solution, and at the same time substitute n for $\sum\limits_{i=1}^{n} 1$:

$$R_n = \frac{27}{n^3} \sum_{i=1}^{n} i^2 + \frac{3}{n} \sum_{i=1}^{n} 1$$

$$= \frac{27}{n^3} \cdot \frac{n(n+1)(2n+1)}{6} + \frac{3}{n} \cdot n$$

$$= \frac{27}{n^3} \cdot \frac{2n^3 + 3n^2 + n}{6} + 3$$

$$= \frac{27}{n^3} \cdot \left(\frac{n^3}{3} + \frac{n^2}{2} + \frac{n}{6} \right) + 3$$

$$= 9 + \frac{27}{2n} + \frac{9}{2n^2} + 3$$

$$= 12 + \frac{27}{2n} + \frac{9}{2n^2}$$

The end. Finally! This is the formula for the area of n right rectangles between $x = 0$ and $x = 3$ under the function $f(x) = x^2 + 1$. You can use this formula to produce the approximate areas given in Table 14-2. But once you've got such a formula, it'd be kind of pointless to produce a table of approximate areas, because you can use the formula to determine the *exact* area. And it's a snap. I get to that in a minute in the next section.

But first, here are the formulas for n left rectangles and n midpoint rectangles between $x=0$ and $x=3$ under the same function, x^2+1. These formulas generate the area approximations in Tables 14-1 and 14-3. The algebra for deriving these formulas is even worse than what you just did for the right rectangle formula, so I decided to skip it. Do you mind? I didn't think so.

$$L_n = 12 - \frac{27}{2n} + \frac{9}{2n^2}$$

$$M_n = 12 - \frac{9}{4n^2}$$

And now, what you've all been waiting for . . .

Finding Exact Area with the Definite Integral

Having laid all the necessary groundwork, you're finally ready to move on to determining exact areas — which is the whole point of integration. You don't need calculus to do all the approximation stuff you just did.

As you saw with the left, right, and midpoint rectangles in the "Approximating Area" sections, the more rectangles you use, the better the approximation. So, "all" you'd have to do to get the exact area under a curve is use an infinite number of rectangles. Now, you can't really do that, but with the fantastic invention of limits, this is sort of what happens. Here's the definition of the definite integral that's used to compute exact areas.

The *definite integral* ("simple" definition): The exact area under a curve between $x=a$ and $x=b$ is given by the definite integral, which is defined as the limit of a Riemann sum:

$$\int_a^b f(x)\,dx = \lim_{n\to\infty} \sum_{i=1}^{n} \left[f(x_i) \cdot \frac{b-a}{n} \right]$$

Is that a thing of beauty or what? The summation above (everything to the right of "lim") is identical to the formula for n right rectangles, R_n, that I give a few pages back. The only difference here is that you take the limit of that formula as the number of rectangles approaches infinity (∞).

This definition of the definite integral is the simple version based on the right rectangle formula. I give you the real-McCoy definition later, but because all Riemann sums for a specific problem have the same limit — in other words, it doesn't matter what type of rectangles you use — you might as well use the right-rectangle definition. It's the least complicated and it'll always suffice.

Let's have a drum roll. Here, finally, is the exact area under our old friend $f(x) = x^2 + 1$ between $x = 0$ and $x = 3$:

$$\int_0^3 (x^2 + 1)\,dx = \lim_{n \to \infty} \sum_{i=1}^n \left[f(x_i) \cdot \frac{b-a}{n} \right]$$

$$= \lim_{n \to \infty} \left(12 + \frac{27}{2n} + \frac{9}{2n^2} \right)$$

(This is what we got in the "Writing Riemann sums with sigma notation" section after all those steps.)

$$= 12 + \frac{27}{2 \cdot \infty} + \frac{9}{2 \cdot \infty^2}$$

$$= 12 + \frac{27}{\infty} + \frac{9}{\infty}$$

$$= 12 + 0 + 0$$

(Remember, in a limit problem, any number divided by infinity equals zero.)

$$= 12$$

Big surprise.

This result is pretty amazing if you think about it. Using the limit process, you get an *exact* answer of 12 — sort of like 12.00000000 . . . to an infinite number of decimal places — for the area under the smooth, curving function $f(x) = x^2 + 1$, based on the areas of flat-topped rectangles that run along the curve in a jagged, sawtooth fashion. Let me guess — the sheer power of this mathematical beauty is bringing tears to your eyes.

Finding the exact area of 12 by using the limit of a Riemann sum is a lot of work (remember, you first had to determine the formula for n right rectangles). This complicated method of integration is comparable to determining a derivative the hard way by using the formal definition that's based on the difference quotient (see Chapter 9). And just as you stopped using the formal definition of the derivative after you learned the differentiation shortcuts, you won't have to use the formal definition of the definite integral based on a Riemann sum after you learn the shortcut methods in Chapters 15 and 16 — except, that is, on your final exam.

Because the limit of all Riemann sums is the same, the limits at infinity of n left rectangles and n midpoint rectangles — for $f(x) = x^2 + 1$ between $x = 0$ and $x = 3$ — should give us the same result as the limit at infinity of n right rectangles. The expressions after the following limit symbols are the formulas for n left rectangles and n midpoint rectangles that appear at the end of the "Writing Riemann sums with sigma notation" section earlier in the chapter. Here's the left rectangle limit:

$$\int_0^3 (x^2+1)dx = L_\infty = \lim_{n \to \infty} \left(12 - \frac{27}{2n} + \frac{9}{2n^2} \right)$$

$$= 12 - \frac{27}{2 \cdot \infty} + \frac{9}{2 \cdot \infty^2}$$

$$= 12 - \frac{27}{\infty} + \frac{9}{\infty}$$

$$= 12 - 0 + 0$$

$$= 12$$

And here's the midpoint rectangle limit:

$$\int_0^3 (x^2+1)dx = M_\infty = \lim_{n \to \infty} \left(12 - \frac{9}{4n^2} \right)$$

$$= 12 - \frac{9}{4 \cdot \infty^2}$$

$$= 12 - \frac{9}{\infty}$$

$$= 12 - 0$$

$$= 12$$

If you're somewhat incredulous that these limits actually give you the *exact* area under $f(x)=x^2+1$ between 0 and 3, you're not alone. After all, in these limits, as in all limit problems, the arrow-number (∞ in this example) is only *approached*; it's never actually reached. And on top of that, what would it mean to reach infinity? You can't do it. And regardless of how many rectangles you have, you always have that jagged, sawtooth edge. So how can such a method give you the exact area?

Look at it this way. You can tell from Figures 14-4 and 14-5 that the sum of the areas of left rectangles, regardless of their number, will always be an *under-estimate* (this is the case for functions that are increasing over the span in question). And from Figure 14-6, you can see that the sum of the areas of right rectangles, regardless of how many you have, will always be an *overestimate* (for increasing functions). So, because the limits at infinity of the underesti-mate and the overestimate are both equal to 12, that must be the exact area. (A similar argument works for decreasing functions.)

All Riemann sums for a given problem have the same limit. Not only are the limits at infinity of left, right, and midpoint rectangles the same for a given problem, the limit of any Riemann sum also gives you the same answer. You can have a series of rectangles with unequal widths; you can have a mix of left, right, and midpoint rectangles; or you can construct the rectangles so they touch the curve somewhere other than at their left or right upper corners or at the midpoints of their top sides. The only thing that matters is that, in the limit, the width of all the rectangles tends to zero (and from this it follows that the number of rectangles approaches infinity). This brings us to the following totally extreme, down-and-dirty integration mumbo jumbo that takes all these possibilities into account.

The *definite integral* (real-McCoy definition): The definite integral from $x = a$ to $x = b$, $\int_a^b f(x)dx$, is the number to which all Riemann sums tend as the width of all rectangles tends to zero and as the number of rectangles approaches infinity:

$$\int_a^b f(x)dx = \lim_{\max \Delta x_i \to 0} \sum_{i=1}^n f(c_i)\Delta x_i,$$

where Δx_i is the width of the ith rectangle and c_i is the x-coordinate of the point where the ith rectangle touches $f(x)$. (That "$\max \Delta x_i \to 0$" simply guarantees that the width of all the rectangles approaches zero and that the number of rectangles approaches infinity.)

Approximating Area with the Trapezoid Rule and Simpson's Rule

This section covers two more ways to estimate the area under a function. You can use them if for some reason you only want an estimate and not an exact answer — maybe because you're asked for that on an exam. But these approximation methods and the others we've gone over are useful for another reason. There are certain types of functions for which the exact area method doesn't work. (It's beyond the scope of this book to explain why this is the case or exactly what these functions are like, so just take my word for it.) So, using an approximation method may be your only choice if you happen to get one of these uncooperative functions.

The trapezoid rule

With the trapezoid rule, instead of approximating area with rectangles, you do it with — can you guess? — trapezoids. See Figure 14-9.

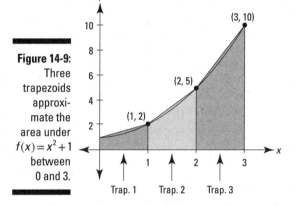

Figure 14-9:
Three trapezoids approximate the area under $f(x) = x^2 + 1$ between 0 and 3.

Because of the way trapezoids hug the curve, they give you a much better area estimate than either left or right rectangles. And it turns out that a trapezoid approximation is the average of the left rectangle and right rectangle approximations. Can you see why? (Hint: The area of a trapezoid — say trapezoid 2 in Figure 14-9 — is the average of the areas of the two corresponding rectangles in the left and right sums, namely, rectangle number 2 in Figure 14-4 and rectangle 2 in Figure 14-6.)

Table 14-4 lists the trapezoid approximations for the area under $f(x) = x^2 + 1$ between $x = 0$ and $x = 3$.

Table 14-4   Estimates of the Area Under $f(x) = x^2 + 1$ between $x = 0$ and $x = 3$ Given by Increasing Numbers of Trapezoids

Number of Trapezoids	Area Estimate
3	12.5
6	12.125
12	~ 12.031
24	~ 12.008
48	~ 12.002
96	~ 12.0005
192	~ 12.0001
384	~ 12.00003

From the look of Figure 14-9, you might expect a trapezoid approximation to be better than a midpoint estimate, but in fact, as a general rule, midpoint estimates are about twice as good as trapezoid estimates. You can confirm this by comparing Tables 14-3 and 14-4. For instance, Table 14-3 lists an area estimate of 11.9990 for 48 midpoint rectangles. This differs from the exact area of 12 by 0.001. The area estimate with 48 trapezoids given in Table 14-4, namely 12.002, differs from 12 by twice as much.

A trapezoid approximation is the average of the corresponding left-rectangle approximation and the right-rectangle approximation. If you've already worked out the left- and right-rectangle approximations for a particular function and a certain number of rectangles, you can just average them to get the corresponding trapezoid estimate. If not, here's the formula:

The trapezoid rule: You can approximate the exact area under a curve between $x = a$ and $x = b$, $\int_a^b f(x)dx$, with a sum of trapezoids given by the following formula. In general, the more trapezoids, the better the estimate.

$$T_n = \frac{b-a}{2n}[f(x_0) + 2f(x_1) + 2f(x_2) + 2f(x_3) + \ldots\ldots + 2f(x_{n-1}) + f(x_n)],$$

where n is the number of trapezoids, $\frac{b-a}{2n}$ is half the "height" of each sideways trapezoid, and x_0 through x_n are the $n+1$ evenly spaced points from $x = a$ to $x = b$. (By the way, using that half-the-height expression is completely unintuitive considering that the formula for the area of a trapezoid uses its height, not half its height. For extra credit, see if you can figure out why that $b - a$ is divided by $2n$ instead of just n.)

Even though the formal definition of the definite integral is based on the sum of an infinite number of *rectangles,* I prefer to think of integration as the limit of the trapezoid rule at infinity. The further you zoom in on a curve, the straighter it gets. When you use a greater and greater number of trapezoids and then zoom in on where the trapezoids touch the curve, the tops of the trapezoids get closer and closer to the curve. If you zoom in "infinitely," the tops of the "infinitely many" trapezoids *become* the curve and, thus, the sum of their areas gives you the exact area under the curve. This is a good way to think about why integration produces the exact area — and it makes sense conceptually — but it's not actually done this way.

Simpson's rule — that's Thomas (1710–1761), not Homer (1987–)

Now I really get fancy and draw shapes that are sort of like trapezoids except that instead of having slanting tops, they have curved, parabolic tops. See Figure 14-10.

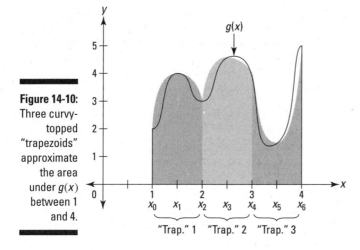

Figure 14-10: Three curvy-topped "trapezoids" approximate the area under $g(x)$ between 1 and 4.

Note that with Simpson's rule each "trapezoid" spans two intervals instead of one; in other words, "trapezoid" number 1 goes from x_0 to x_2, "trapezoid" 2 goes from x_2 to x_4, and so on. Because of this, the total span must always be divided into an even number of intervals.

Simpson's rule is by far the most accurate approximation method discussed in this chapter. In fact, it gives the *exact* area for any polynomial function of degree three or less. In general, Simpson's rule gives a much better estimate than either the midpoint rule or the trapezoid rule.

You can use a midpoint sum with a trapezoid sum to calculate a Simpson sum. A Simpson's rule sum is sort of an average of a midpoint sum and a trapezoid sum, except that you use the midpoint sum twice in the average. So, if you already have the midpoint sum and the trapezoid sum for some number of rectangles/trapezoids, you can obtain the Simpson's rule approximation with the following simple average:

$$S_{2n} = \frac{M_n + M_n + T_n}{3}$$

Note the subscript of $2n$. This means that if you use, say, M_3 and T_3, you get a result for S_6. But S_6, which has six intervals, has only three curvy "trapezoids" because each of them spans two intervals. Thus, the above formula always involves the same number of rectangles, trapezoids, and Simpson's rule "trapezoids."

If you don't have the midpoint and trapezoid sums for the above shortcut, you can use the following formula for Simpson's rule.

Simpson's rule: You can approximate the exact area under a curve between $x = a$ and $x = b$, $\int_a^b f(x)dx$, with a sum of parabola-topped "trapezoids" given by the following formula. In general, the more "trapezoids," the better the estimate.

$$S_n = \frac{b-a}{3n}\left[f(x_0) + 4f(x_1) + 2f(x_2) + 4f(x_3) + 2f(x_4) + \ldots + 4f(x_{n-1}) + f(x_n)\right],$$

where n is twice the number of "trapezoids" and x_0 through x_n are the $n+1$ evenly spaced points from $x = a$ to $x = b$.

To close this chapter, here's a warning about functions that go below the x-axis. I didn't include any such functions in this chapter, because I thought you already had enough to deal with. You see the full explanation and an example in Chapter 17.

Areas *below* the x-axis count as *negative* areas. Whether approximating areas with right-, left-, or midpoint rectangles or with the trapezoid rule or Simpson's rule, or computing exact areas with the definite integral, areas below the x-axis and above the curve count as *negative* areas.

Chapter 15

Integration: It's Backwards Differentiation

 C hapter 14 shows you the hard way to calculate the area under a function using the formal definition of integration — the limit of a Riemann sum. In this chapter, I calculate areas the easy way, taking advantage of one of the most important and amazing discoveries in mathematics — that integration (finding areas) is just differentiation in reverse. That reverse process was a great discovery, and it's based on some difficult ideas, but before we get to that, let's talk about a related, straightforward reverse process, namely. . . .

Antidifferentiation

The derivative of sin$x$ is cos$x$, so the antiderivative of cos$x$ is sin$x$; the derivative of x^3 is $3x^2$, so the antiderivative of $3x^2$ is x^3 — you just go backwards. There's a bit more to it, but that's the basic idea. Later in this chapter, I show you how to find areas by using antiderivatives. This is *much* easier than finding areas with the Riemann sum technique.

Now consider x^3 and its derivative $3x^2$ again. The derivative of $x^3 + 10$ is also $3x^2$, as is the derivative of $x^3 - 5$. Any function of the form $x^3 + C$, where C is any number, has a derivative of $3x^2$. So, every such function is an antiderivative of $3x^2$.

Definition of the *indefinite integral:* The indefinite integral of a function $f(x)$, written as $\int f(x)\,dx$, is the family of *all* antiderivatives of the function. For example, because the derivative of x^3 is $3x^2$, the indefinite integral of $3x^2$ is $x^3 + C$, and you write

$$\int 3x^2 dx = x^3 + C$$

You probably recognize this integration symbol, $\int$, from the discussion of the *definite* integral in Chapter 14. The definite integral symbol, however, contains two little numbers like $\int_{4}^{10}$ that tell you to compute the area under a function between those two numbers, called the *limits of integration.* The naked version of the symbol, $\int$, indicates an *indefinite* integral or an *antiderivative.* This chapter is all about the intimate connection between these two symbols, these two ideas.

Figure 15-1 shows the family of antiderivatives of $3x^2$, namely $x^3 + C$. Note that this family of curves has an infinite number of curves. They go up and down forever and are infinitely dense. The vertical gap of 2 units between each curve in Figure 15-1 is just a visual aid.

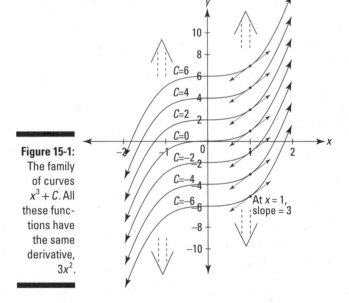

Figure 15-1:
The family of curves $x^3 + C$. All these functions have the same derivative, $3x^2$.

Consider a few things about Figure 15-1. The top curve on the graph is $y = x^3 + 6$; the one below it is $y = x^3 + 4$; the bottom one $y = x^3 - 6$. By the power rule, these three functions, as well as all the others in this family of functions, have a derivative of $3x^2$. Now, consider the slope of each of the curves where x equals 1 (see the tangent lines drawn on the curves). The derivative of each function is $3x^2$, so when x equals 1, the slope of each curve is $3 \cdot 1^2$, or 3. Thus, all these little tangent lines are parallel. Next, notice that all the functions in Figure 15-1 are identical except for being slid up or down (remember vertical shifts from Chapter 5?). Because they differ only by a vertical shift, the steepness at any x-value, like at $x = 1$, is the same for all the curves. This is the visual way to understand why each of these curves has the same derivative, and, thus, why each curve is an antiderivative of the same function.

Vocabulary, Voshmabulary: What Difference Does It Make?

In general, definitions and vocabulary are very important in mathematics, and it's a good idea to use them correctly. But with the current topic, I'm going to be a bit lazy about precise terminology, and I hereby give you permission to do so as well.

If you're a stickler, you should say that *the* indefinite integral of $3x^2$ is $x^3 + C$ and that $x^3 + C$ is the family or set of *all* antiderivatives of $3x^2$ (you don't say that $x^3 + C$ is *the* antiderivative), and you say that $x^3 + 10$, for instance, is *an* antiderivative of $3x^2$. And on a test, you should definitely write $\int 3x^2 dx = x^3 + C$. If you leave the C off, you'll likely lose some points.

But, when discussing these matters, no one will care or be confused if you get tired of saying "+ C" after every indefinite integral and just say, for example, that the indefinite integral of $3x^2$ is x^3, and you can skip the *indefinite* and just say that the *integral* of $3x^2$ is x^3. And instead of always talking about that family of functions business, you can just say that *the* antiderivative of $3x^2$ is $x^3 + C$ or that the antiderivative of $3x^2$ is x^3. Everyone will know what you mean. It may cost me my membership in the National Council of Teachers of Mathematics, but at least occasionally, I use this loose approach.

The Annoying Area Function

This is a tough one — gird your loins. Say you've got any old function, $f(t)$. Imagine that at some t-value, call it s, you draw a fixed vertical line. See Figure 15-2.

Then you take a moveable vertical line, starting at the same point, s ("s" is for *starting* point), and drag it to the right. As you drag the line, you sweep out a larger and larger area under the curve. This area is a function of x, the position of the moving line. In symbols, you write

$$A_f(x) = \int_s^x f(t)\, dt$$

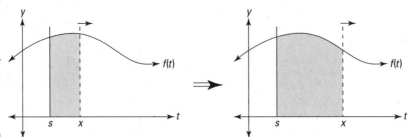

Note that *t* is the input variable in $f(t)$ instead of *x* because *x* is already taken — it's the input variable in $A_f(x)$. The subscript *f* in A_f indicates that $A_f(x)$ is the area function for the particular curve *f* or $f(t)$. The *dt* is a little increment along the *t*-axis — actually an infinitesimally small increment.

Here's a simple example to make sure you've got a handle on how an area function works. By the way, don't feel bad if you find this extremely hard to grasp — you've got lots of company. Say you've got the simple function, $f(t) = 10$, that's a horizontal line at $y = 10$. If you sweep out area beginning at $s = 3$, you get the following area function:

$$A_f(x) = \int_3^x 10\, dt$$

You can see that the area swept out from 3 to 4 is 10 because, in dragging the line from 3 to 4, you sweep out a rectangle with a width of 1 and a height of 10, which has an area of 1 times 10, or 10. See Figure 15-3.

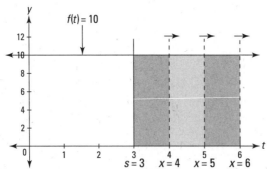

So, $A_f(4)$, the area swept out as you hit 4, equals 10. $A_f(5)$ equals 20 because when you drag the line to 5, you've swept out a rectangle with a width of 2 and height of 10, which has an area of 2 times 10, or 20. $A_f(6)$ equals 30, and so on.

Now, imagine that you drag the line across at a rate of one unit per second. You start at $x = 3$, and you hit 4 at 1 second, 5 at 2 seconds, 6 at 3 seconds, and so on. How much area are you sweeping out per second? Ten square units per second because each second you sweep out another 1-by-10 rectangle. Notice — this is huge — that because the width of each rectangle you sweep out is 1, the area of each rectangle — which is given by *height* times *width* — is the same as its height because anything times 1 equals itself. You see why this is huge in a minute. (By the way, the real rate we care about here is not area swept out per second, but, rather, area swept out per unit change on the *x*-axis. I explain it in terms of per second because it's easier to think about a sweeping-out-area rate this way. And since you're dragging the line across at *one* *x*-axis unit per *one* second, both rates are the same. Take your pick.)

The derivative of an area function equals the rate of area being swept out. Okay, are you sitting down? You've reached another one of the big *Ah ha!* moments in the history of mathematics. Recall that *a derivative is a rate*. So, because the rate at which the previous area function grows is 10 square units per second, you can say its derivative equals 10. Thus, you can write

$$\frac{d}{dx} A_f(x) = 10$$

Again, this just tells you that with each 1 unit increase in x, A_f (the area function) goes up 10. Now here's the critical thing: Notice that this rate or derivative of 10 is the same as the height of the original function $f(t) = 10$ because as you go across 1 unit, you sweep out a rectangle that's 1 by 10, which has an area of 10, the height of the function.

And the rate works out to 10 regardless of the width of the rectangle. Imagine that you drag the vertical line from $x = 4$ to $x = 4.001$. At a rate of one unit per second, that'll take you 1/1000th of a second, and you'll sweep out a skinny rectangle with a width of 1/1000, a height of 10, and thus an area of 10 times 1/1000, or 1/100 square units. The rate of area being swept out would be, therefore, $\dfrac{1/100 \text{ square units}}{1/1000 \text{ seconds}}$ which equals 10 square units per second.

So you see that with every small increment along the *x*-axis, the rate of area being swept out equals the function's height.

This works for any function, not just horizontal lines. Look at the function $g(t)$ and its area function $A_g(x)$ that sweeps out area beginning at $s = 2$ in Figure 15-4.

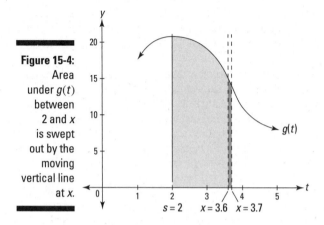

Between $x = 3.6$ and $x = 3.7$, $A_g(x)$ grows by the area of that skinny, dark shaded "rectangle" with a width of 0.1 and a height of about 15. (As you can see, it's not really a rectangle; it's closer to a trapezoid, but it's not that either because its tiny top is curving slightly. But, in the limit, as the width gets smaller and smaller, the skinny "rectangle" behaves precisely like a real rectangle.) So, to repeat, $A_g(x)$ grows by the area of that dark "rectangle" which has an area extremely close to 0.1 times 15, or 1.5. That area is swept out in 0.1 seconds, so the rate of area being swept out is $\dfrac{1.5 \text{ square units}}{0.1 \text{ seconds}}$, or 15 square units per second, the height of the function. This idea is so important that it deserves an icon. . . .

The sweeping out area rate equals the height. The *rate* of area being swept out under a curve by an area function at a given x-value is equal to the *height* of the curve at that x-value.

The Power and the Glory of the Fundamental Theorem of Calculus

Sound the trumpets! Now that you've seen the connection between the rate of growth of an area function and the height of the given curve, you're ready for the fundamental theorem of calculus — what some say is one of the most important theorems in the history of mathematics.

The fundamental theorem of calculus: Given an area function A_f that sweeps out area under $f(t)$,

$$A_f(x) = \int_s^x f(t)\, dt,$$

the rate at which area is being swept out is equal to the height of the original function. So, because the rate is the derivative, the derivative of the area function equals the original function:

$$\frac{d}{dx} A_f(x) = f(x).$$

Because $A_f(x) = \int_s^x f(t)\, dt$, you can also write the above equation as follows:

$$\frac{d}{dx} \int_s^x f(t)\, dt = f(x)$$

Break out the smelling salts.

Now, because the derivative of $A_f(x)$ is $f(x)$, $A_f(x)$ is by definition an *antiderivative* of $f(x)$. Check out how this works by returning to the simple function from the previous section, $f(t) = 10$, and its area function, $A_f(x) = \int_s^x 10\, dt$.

According to the fundamental theorem, $\frac{d}{dx} A_f(x) = 10$. Thus A_f must be an antiderivative of 10; in other words, A_f is a function whose derivative is 10. Because any function of the form $10x + C$, where C is a number, has a derivative of 10, the antiderivative of 10 is $10x + C$. The particular number C depends on your choice of s, the point where you start sweeping out area. For a particular choice of s, the area function will be the one function (out of all the functions in the family of curves $10x + C$) that crosses the x-axis at s. To figure out C, set the antiderivative equal to zero, plug the value of s into x, and solve for C.

For this function with an antiderivative of $10x + C$, if you start sweeping out area at, say, $s = 0$, then $10 \cdot 0 + C = 0$, so $C = 0$, and thus, $A_f(x) = \int_0^x 10\, dt = 10x + C$, or just $10x$. (Note that C does not necessarily equal s. In fact, it usually doesn't (especially when $s \neq 0$). When $s = 0$, C often also equals 0, but not for all functions.)

Figure 15-5 shows why $A_f(x) = 10x$ is the correct area function if you start sweeping out area at zero. In the top graph in the figure, the area under the curve from 0 to 3 is 30, and that's given by $A_f(3) = 10 \cdot 3 = 30$. And you can see that the area from 0 to 5 is 50, which agrees with the fact that $A_f(5) = 10 \cdot 5 = 50$.

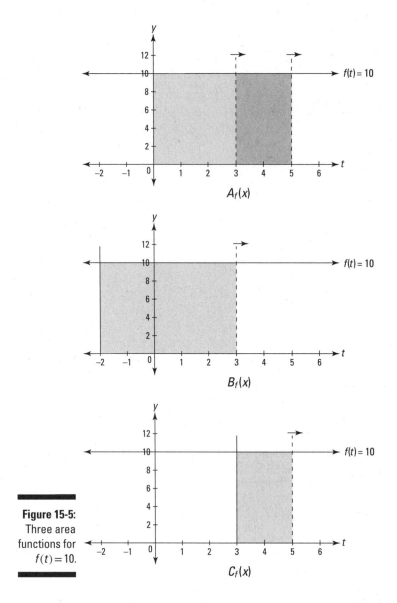

If instead you start sweeping out area at $s = -2$ and define a new area function, $B_f(x) = \int_{-2}^{x} 10\,dt$, then $10 \cdot (-2) + C = 0$, so C equals 20 and $B_f(x)$ is thus $10x + 20$.

This area function is 20 more than $A_f(x)$, which starts at $s = 0$, because if you start at $s = -2$, you've already swept out an area of 20 by the time you get to zero. Figure 15-5 shows why $B_f(3)$ is 20 more than $A_f(3)$.

And if you start sweeping out area at $s = 3$, $10 \cdot 3 + C = 0$, so $C = -30$ and the

area function is $C_f(x) = \int_3^x 10 \, dt = 10x - 30$. This function is 30 *less* than $A_f(x)$

because with $C_f(x)$, you lose the 3-by-10 rectangle between 0 and 3 that $A_f(x)$ has (see the bottom graph in Figure 15-5).

An area function is an antiderivative. The area swept out under the horizontal line $f(t) = 10$, from some number s to x, is given by an antiderivative of 10, namely $10x + C$, where the value of C depends on where you start sweeping out area.

Now let's look at graphs of $A_f(x)$, $B_f(x)$, and $C_f(x)$. (Note that Figure 15-5 doesn't show the graphs of $A_f(x)$, $B_f(x)$, and $C_f(x)$. You see three graphs of the horizontal line function, $f(t) = 10$; and you see the areas swept out under $f(t)$ by $A_f(x)$, $B_f(x)$, and $C_f(x)$, but you don't actually see the graphs of these three area functions.) Check out Figure 15-6.

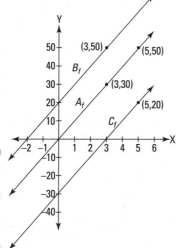

Figure 15-6: The actual graphs of $A_f(x)$, $B_f(x)$, and $C_f(x)$.

Figure 15-6 shows the graphs of the equations of $A_f(x)$, $B_f(x)$, and $C_f(x)$ which we worked out before: $A_f(x) = 10x$, $B_f(x) = 10x + 20$, and $C_f(x) = 10x - 30$. (As you can see, all three are simple, $y = mx + b$ lines.) The y-values of these three functions give you the areas swept out under $f(t) = 10$ that you see in Figure 15-5. Note that the three x-intercepts you see in Figure 15-6 are the three x-values in Figure 15-5 where sweeping out area begins.

We worked out above that $A_f(3) = 30$ and that $A_f(5) = 50$. You can see those areas of 30 and 50 in the top graph of Figure 15-5. In Figure 15-6, you see these results on A_f at the points $(3, 30)$ and $(5, 50)$. You also saw in Figure 15-5 that

$B_f(3)$ was 20 more than $A_f(3)$; you see that result in Figure 15-6 where $(3, 50)$ on B_f is 20 higher than $(3, 30)$ on A_f. Finally, you saw in Figure 15-5 that $C_f(x)$ is 30 less than $A_f(x)$. Figure 15-6 shows that in a different way: at any x-value, the C_f line is 30 units below the A_f line.

A few observations. You already know from the fundamental theorem that $\frac{d}{dx}A_f(x)=f(x)=10$ (and the same for $B_f(x)$ and $C_f(x)$). That was explained above in terms of rates: For A_f, B_f, and C_f, the rate of area being swept out under $f(t)=10$ equals 10. Figure 15-6 also shows that $\frac{d}{dx}A_f(x)=10$ (and the same for B_f and C_f), but here you see the derivative as a slope. The slopes, of course, of all three lines equal 10. Finally, note that — like you saw in Figure 15-1 — the three lines in Figure 15-6 differ from each other only by a vertical translation. These three lines (and the infinity of all other vertically translated lines) are all members of the class of functions, $10x+C$, the family of antiderivatives of $f(x)=10$.

For the next example, look again at the parabola $y=x^2+1$, our friend from Chapter 14 which we analyzed in terms of the sum of the areas of rectangles (Riemann sums). Flip back to Figure 14-4, and check out the shaded region under $y=x^2+1$. Now you can finally compute the exact area of the shaded region the easy way.

The area function for sweeping out area under x^2+1 is $A_f(x)=\int_s^x(t^2+1)\,dt$. By the fundamental theorem, $\frac{d}{dx}A_f(x)=x^2+1$, and so A_f is an antiderivative of x^2+1. Any function of the form $\frac{1}{3}x^3+x+C$ has a derivative of x^2+1 (try it), so that's the antiderivative. For Figure 14-6, you want to sweep out area beginning at 0, so $s=0$. Set the antiderivative equal to zero, plug the value of s into x, and solve for C: $\frac{1}{3}\cdot0^3+0+C=0$, so $C=0$, and thus

$$A_f(x)=\int_0^x(t^2+1)\,dt=\frac{1}{3}x^3+x+0$$

The area swept out from 0 to 3 — which we did the hard way in Chapter 14 by computing the limit of a Riemann sum — is simply $A_f(3)$:

$$A_f(x)=\frac{1}{3}x^3+x$$

$$A_f(3)=\frac{1}{3}\cdot3^3+3=9+3=12$$

Piece o' cake. That was *much* less work than doing it the hard way.

And after you know that the area function that starts at zero, $\int_0^x (t^2+1)\,dt$, equals $\frac{1}{3}x^3+x$, it's a snap to figure the area of other sections under the parabola that don't start at zero. Say, for example, you want the area under the parabola between 2 and 3. You can compute that area by subtracting the area between 0 and 2 from the area between 0 and 3. You just figured the area between 0 and 3 — that's 12. And the area between 0 and 2 is $A_f(2)=\frac{1}{3}\cdot 2^3+2=4\frac{2}{3}$. So the area between 2 and 3 is $12-4\frac{2}{3}$, or $7\frac{1}{3}$. This subtraction method brings us to the next topic — the second version of the fundamental theorem.

The Fundamental Theorem of Calculus: Take Two

Now we finally arrive at the super-duper shortcut integration theorem that you'll use for the rest of your natural born days — or at least till the end of your stint with calculus. This shortcut method is all you need for the integration word problems in Chapters 17 and 18.

The fundamental theorem of calculus (second version or shortcut version): Let F be any antiderivative of the function f; then

$$\int_a^b f(x)\,dx = F(b)-F(a)$$

This theorem gives you the super shortcut for computing a definite integral like $\int_2^3 (x^2+1)\,dx$, the area under the parabola $y=x^2+1$ between 2 and 3. As I show in the previous section, you can get this area by subtracting the area between 0 and 2 from the area between 0 and 3, but to do that you need to know that the particular area function sweeping out area beginning at zero, $\int_0^x (t^2+1)\,dt$, is $\frac{1}{3}x^3+x$ (with a C value of zero).

The beauty of the shortcut theorem is that you don't have to even use an area function like $A_f(x)=\int_0^x (t^2+1)\,dt$. You just find any antiderivative, $F(x)$, of your function, and do the subtraction, $F(b)-F(a)$. The simplest antiderivative to use is the one where $C = 0$. So here's how you use the theorem to find the area under our parabola from 2 to 3. $F(x) = \frac{1}{3}x^3 + x$ is an antiderivative of $x^2 + 1$. Then the theorem gives you:

$$\int_2^3 (x^2 + 1)dx = F(3) - F(2)$$

$F(3) - F(2)$ can be written as $\left[\frac{1}{3}x^3 + x\right]_2^3$, and thus,

$$\int_2^3 (x^2 + 1)dx = \left[\frac{1}{3}x^3 + x\right]_2^3$$
$$= \left(\frac{1}{3} \cdot 3^3 + 3\right) - \left(\frac{1}{3} \cdot 2^3 + 2\right)$$
$$= 12 - 4\frac{2}{3}$$
$$= 7\frac{1}{3}$$

Granted, this is the same computation I did in the previous section using the area function with $s = 0$, but that's only because for the $y = x^2 + 1$ function, when s is zero, C is also zero. It's sort of a coincidence, and it's not true for all functions. But regardless of the function, the shortcut works, and you don't have to worry about area functions or s or C. All you do is $F(b) - F(a)$.

Here's another example: What's the area under $f(x) = e^x$ between $x = 3$ and $x = 5$? The derivative of e^x is e^x, so e^x is an antiderivative of e^x, and thus

$$\int_3^5 e^x dx = \left[e^x\right]_3^5$$
$$= e^5 - e^3$$
$$\approx 148.4 - 20.1$$
$$\approx 128.3$$

What could be simpler?

Areas *above* the curve and *below* the x-axis count as *negative* areas. Before going on, I'd be remiss if I didn't touch on negative areas (this is virtually the same caution made at the very end of Chapter 14). Note that with the two examples above, the parabola, $y = x^2 + 1$, and the exponential function, $y = e^x$, the areas we're computing are *under* the curves and *above* the x-axis. These areas count as ordinary, *positive* areas. But, if a function goes below the *x*-axis, areas above the curve and below the *x*-axis count as *negative* areas. This is the case whether you're using an area function, the first version of the fundamental theorem of calculus, or the shortcut version. Don't worry about this for now. You see how this works in Chapter 17.

Okay, so now you've got the super shortcut for computing the area under a curve. And if one big shortcut wasn't enough to make your day, Table 15-1 lists some rules about definite integrals that can make your life much easier.

Table 15-1	Five Easy Rules for Definite Integrals

1) $\displaystyle\int_a^a f(x)\,dx = 0$ (Well, duh – there's no area "between" a and a.)

2) $\displaystyle\int_b^a f(x)\,dx = -\int_a^b f(x)\,dx$

3) $\displaystyle\int_a^b f(x)\,dx = \int_a^c f(x)\,dx + \int_c^b f(x)\,dx$

4) $\displaystyle\int_a^b kf(x)\,dx = k\int_a^b f(x)\,dx$ (k is a constant; you can pull a constant out of the integral.)

5) $\displaystyle\int_a^b [f(x) \pm g(x)]\,dx = \int_a^b f(x)\,dx \pm \int_a^b g(x)\,dx$

Now that I've given you the shortcut version of the fundamental theorem, that doesn't mean you're off the hook. Below are three different ways to understand why the theorem works. This is difficult stuff — brace yourself.

Alternatively, you can skip these explanations if all you want to know is how to compute an area: forget about *C* and just subtract $F(a)$ from $F(b)$. I include these explanations because I suspect you're dying to learn extra math just for the love of learning — right? Other books just give you the rules; I explain why they work and the underlying principles — that's why they pay me the big bucks.

Actually, in all seriousness, you should read at least some of this material. The fundamental theorem of calculus is one of the most important theorems in all of mathematics, so you ought to spend some time trying hard to understand what it's all about. It's worth the effort. Of the three explanations, the first is the easiest. But if you only want to read one or two of the three, I'd read just the third, or the second and the third. Or, you could begin with the figures accompanying the three explanations, because the figures really show you what's going on. Finally, if you can't digest all of this in one sitting — no worries — you can revisit it later.

Why the theorem works: Area functions explanation

One way to understand the shortcut version of the fundamental theorem is by looking at area functions. As you can see in Figure 15-7, the dark-shaded area between a and b can be figured by starting with the area between s and b, then cutting away (subtracting) the area between s and a. And it doesn't matter whether you use 0 as the left edge of the areas or any other value of s. Do you see that you'd get the same result whether you use the graph on the left or the graph on the right?

Figure 15-7: Figuring the area between a and b with two different area functions.

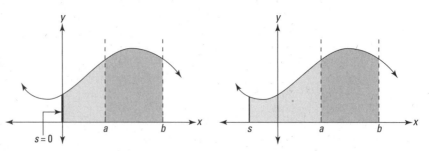

Take a look at $f(t) = 10$ (see Figure 15-8). Say you want the area between 5 and 8 under the horizontal line $f(t) = 10$, and you are forced to use calculus.

Figure 15-8: The shaded area equals 30 — well, duh, it's a 3-by-10 rectangle.

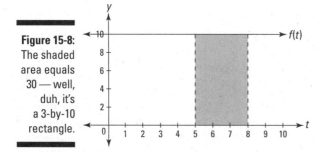

Look back at two of the area functions for $f(t) = 10$ in Figure 15-5: $A_f(x)$ starting at 0 (in which $C = 0$) and $B_f(x)$ starting at -2 (where $C = 20$):

$$A_f(x) = \int_0^x 10\, dt = 10x$$

$$B_f(x) = \int_{-2}^x 10\, dt = 10x + 20$$

If you use $A_f(x)$ to compute the area between 5 and 8 in Figure 15-8, you get the following:

$$\int_5^8 10\, dx = A_f(8) - A_f(5)$$

$$= 10 \cdot 8 - 10 \cdot 5$$

$$= 80 - 50 \qquad \text{(80 is the area of the rectangle from 0 to 8;}$$
$$\text{50 is the area of the rectangle from 0 to 5.)}$$

$$= 30$$

If, on the other hand, you use $B_f(x)$ to compute the same area, you get the same result:

$$\int_5^8 10\, dx = B_f(8) - A_f(5)$$

$$= (10 \cdot 8 + 20) - (10 \cdot 5 + 20)$$

$$\text{(This is } 100 - 70\text{, of course;}$$
$$= (80 + 20) - (50 + 20) \qquad \text{100 is the area of the rectangle from } -2 \text{ to 8;}$$
$$\text{70 is the area of the rectangle from } -2 \text{ to 5.)}$$

$$= 30$$

Notice that the two 20s in the second line from the bottom cancel. Recall that all antiderivatives of $f(t) = 10$ are of the form $10x + C$. Regardless of the value of C, it cancels out as in this example. Thus, you can use any antiderivative with any value of C. For convenience, everyone just uses the antiderivative with $C = 0$, so that you don't mess with C at all. And the choice of s (the point where the area function begins) is irrelevant. So when you're using the shortcut version of the fundamental theorem, and computing an area with $F(b) - F(a)$, you're sort of using a mystery area function with a C value of zero and an unknown starting point, s. Get it?

Why the theorem works: The integration-differentiation connection

The next explanation of the shortcut version of the fundamental theorem involves the yin/yang relationship between differentiation and integration. Check out Figure 15-9.

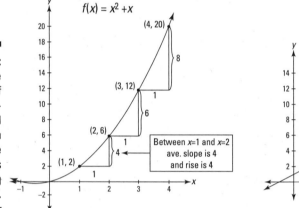

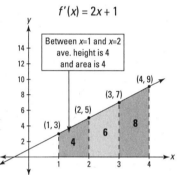

The figure shows a function, $f(x) = x^2 + x$, and its derivative, $f'(x) = 2x + 1$. Look carefully at the numbers 4, 6, and 8 on both graphs. The connection between 4, 6, and 8 on the graph of f — which are the amounts of *rise* between consecutive points on the curve — and 4, 6, and 8 on the graph of f' — which are the *areas* of the trapezoids under f' — shows the intimate relationship between integration and differentiation. Figure 15-9 is a picture worth a thousand symbols and equations, encapsulating the essence of integration in a single snapshot. It shows how the shortcut version of the fundamental theorem works because it shows that the *area* under $f'(x)$ between 1 and 4 equals the total *rise* on $f(x)$ between $(1, 2)$ and $(4, 20)$, in other words that

$$\int_1^4 f'(x) = f(4) - f(1)$$

Note that I've called the two functions in Figure 15-9 and in the above equation f and f' to emphasize that $2x + 1$ is the *derivative* of $x^2 + x$. I could have instead referred to $x^2 + x$ as F and referred to $2x + 1$ as f which would emphasize that $x^2 + x$ is an *antiderivative* of $2x + 1$. In that case you would write the above area equation in the standard way,

$$\int_{1}^{4} f(x)dx = F(4) - F(1)$$

Either way, the meaning's the same. I use the derivative version to point out how finding area is differentiation in reverse. Going from left to right in Figure 15-9 is differentiation: The slopes of *f* correspond to heights on *f'*. Going from right to left is integration: Areas under *f'* correspond to the change in height between two points on *f*.

Okay, here's how it works. Imagine you're going up along *f* from $(1, 2)$ to $(2, 6)$. Every point along the way has a certain steepness, a slope. This slope is plotted as the *y*-coordinate, or height, on the graph of *f'*. The fact that *f'* goes up from $(1, 3)$ to $(2, 5)$ tells you that the slope of *f* goes up from 3 to 5 as you travel between $(1, 2)$ and $(2, 6)$. This all follows from basic differentiation.

Now, as you go along *f* from $(1, 2)$ to $(2, 6)$, the slope is constantly changing. But it turns out that because you go up a total *rise* of 4 as you *run* across 1, the average of all the slopes on *f* between $(1, 2)$ and $(2, 6)$ is $\frac{4}{1}$, or 4. Because each of these slopes is plotted as a *y*-coordinate or height on *f'*, it follows that the average height of *f'* between $(1, 3)$ and $(2, 5)$ is also 4. Thus, between two given points, average slope on *f* equals average height on *f'*.

Hold on, you're almost there. *Slope* equals $\frac{rise}{run}$, so when the run is 1, the slope equals the rise. For example, from $(1, 2)$ to $(2, 6)$ on *f*, the curve rises up 4 and the average slope between those points is also 4. Thus, between any two points on *f* whose *x*-coordinates differ by 1, the average slope *is* the rise.

The area of a trapezoid like the ones on the right in Figure 15-9 equals its width times its average height. (This is true of any other similar shape that has a bottom like a rectangle; the top can be any crooked line or funky curve you like.) So, because the width of each trapezoid is 1, and because anything times 1 is itself, the average height of each trapezoid under *f'* *is* its area; for instance, the area of that first trapezoid is 4 and its average height is also 4.

Are you ready for the grand finale? Here's the whole argument in a nutshell. On *f*, *rise = average slope*; going from *f* to *f'*, *average slope = average height*; on *f'*, *average height = area*. So that gives you *rise = slope = height = area*, and thus, finally, *rise = area*. And that's what the second version of the fundamental theorem says:

$$f(b) - f(a) = \int_{a}^{b} f'(x)dx$$

$$rise = area$$

These ideas are unavoidably difficult. You may have to read it two or three times for it to really sink in.

Notice that it makes no difference to the relationship between slope and area if you use any other function of the form $x^2 + x + C$ instead of $x^2 + x$. Any parabola like $x^2 + x + 10$ or $x^2 + x - 5$ is exactly the same shape as $x^2 + x$; it's just been slid up or down vertically. Any such parabola rises up between $x = 1$ and $x = 4$ in precisely the same way as the parabola in Figure 15-9. From 1 to 2 these parabolas go over 1, up 4. From 2 to 3 they go over 1, up 6, and so on. This is why any antiderivative can be used to find area. The total *area* under f' between 1 to 4, namely 18, corresponds to the total *rise* on any of these parabolas from 1 to 4, namely $4 + 6 + 8$, or 18.

At the risk of beating a dead horse, I've got a third explanation of the fundamental theorem for you. You might prefer it to the first two because it's less abstract — it's connected to simple, commonsense ideas encountered in our day-to-day world. This explanation has a lot in common with the previous one, but the ideas are presented from a different angle.

Why the theorem works: A connection to — egad! — statistics

Don't let the title of this section put you off. I realize that many readers of this calculus book may not have studied statistics. No worries; the statistics connection I explain below involves a very simple thing covered in statistics courses, but you don't need to know any statistics at all to understand this idea. The simple idea is the relationship between a frequency distribution graph and a cumulative frequency distribution graph (you may have run across such graphs in a newspaper or magazine). Consider Figure 15-10.

The upper graph in the figure shows a frequency distribution histogram of the annual profits of Widgets-R-Us from January 1, 2001 through December 31, 2013. The rectangle marked '07, for example, shows that the company's profit for 2007 was $2,000,000 (their best year during the period 2001–2013).

The lower graph in the figure is a cumulative frequency distribution histogram for the same data used for the upper graph. The difference is simply that in the cumulative graph, the height of each column shows the total profits earned since 1/1/2001. Look at the '02 column in the lower graph and the '01 and '02 rectangles in the upper graph, for example. You can see that the '02 column shows the '02 rectangle sitting on top of the '01 rectangle which gives that '02 column a height equal to the total of the profits from '01 and '02. Got it? As you go to the right on the cumulative graph, the height of each successive column simply grows by the amount of profits earned in the corresponding single year shown in the upper graph.

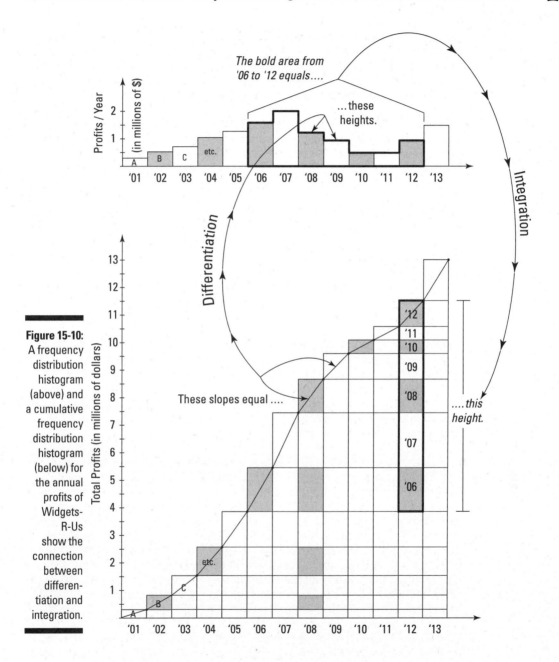

Figure 15-10:
A frequency distribution histogram (above) and a cumulative frequency distribution histogram (below) for the annual profits of Widgets-R-Us show the connection between differentiation and integration.

Okay. So here's the calculus connection. (Bear with me; it takes a while to walk through all this.) Look at the top rectangle of the '08 column on the cumulative graph (let's call that graph *C* for short). At that point on *C*, you *run* across 1 year and *rise* up $1,250,000, the '08 profit you see on the frequency distribution graph (*F* for short). *Slope = rise/run*, so, since the run equals 1, the slope equals 1,250,000/1, or just 1,250,000, which is, of course, the same as the rise. Thus, the slope on C (at '08 or any other year) can be read as a height on F for the corresponding year. (Make sure you see how this works.) Since the heights (or function values) on F are the *slopes* of C, F is the *derivative* of C. In short, F, the derivative, tells us about the slope of C.

The next idea is that since F is the derivative of C, C, by definition, is the antiderivative of F (for example, C might equal $5x^3$ and F would equal $15x^2$). Now, what does C, the antiderivative of F, tell us about F? Imagine dragging a vertical line from left to right over F. As you sweep over the rectangles on F — year by year — the total profit you're sweeping over is shown climbing up along C.

Look at the '01 through '08 rectangles on F. You can see those same rectangles climbing up stair-step fashion along C (see the rectangles labeled A, B, C, etc. on both graphs). The heights of the rectangles from F keep adding up on C as you climb up the stair-step shape. And I've shown how the same '01 through '08 rectangles that lie along the stair-step top of C can also be seen in a vertical stack at year '08 on C. I've drawn the cumulative graph his way so it's even more obvious how the heights of the rectangles add up. (***Note:*** Most cumulative histograms are not drawn this way.)

Each rectangle on F has a base of 1 year, so, since *area = base × height*, the area of each rectangle equals its height. So, as you stack up rectangles on C, you're adding up the areas of those rectangles from F. For example, the height of the '01 through '08 stack of rectangles on C ($8.5 million) equals the total area of the '01 through '08 rectangles on F. And, therefore, the heights or function values of C — which is the antiderivative of F — give you the area under the top edge of F. That's how integration works.

Okay, we're just about done. Now let's go through how these two graphs explain the shortcut version of the fundamental theorem of calculus and the relationship between differentiation and integration. Look at the '06 through '12 rectangles on F (with the bold border). You can see those same rectangles in the bold portion of the '12 column of C. The height of that bold stack, which shows the total profits made during those 7 years, $7.75 million, equals the total area of the 7 rectangles in F. And to get the height of that stack on C, you simply subtract the height of the stack's bottom edge from the height of its upper edge. That's really all the shortcut version of the fundamental theorem says: The *area* under any portion of a function (like F) is given by the change in *height* on the function's antiderivative (like C).

In a nutshell (keep looking at those rectangles with the bold border in both graphs), the *slopes* of the rectangles on C appear as *heights* on F. That's *differentiation*. Reversing direction, you see *integration*: the change in *heights* on C shows the *area* under F. Voilà: differentiation and integration are two sides of the same coin.

(*Note:* Mathematical purists may object to this explanation of the fundamental theorem because it involves *discrete* graphs (for example, the fact that the cumulative distribution histogram in Figure 15-10 goes up at one-year increments), whereas calculus is the study of smooth, continuously changing graphs (the calculus version of the cumulative distribution histogram would be a smooth curve that would show the total profits growing every millisecond— actually, in theory, every infinitesimal fraction of a second). Okay — objection noted — but the fact is that the explanation here does accurately show how integration and differentiation are related and does correctly show how the shortcut version of the fundamental theorem works. All that's needed to turn Figure 15-10 and the accompanying explanation into standard calculus is to take everything to the limit, making the profit interval shorter and shorter and shorter: from a year to a month to a day, etc., etc. In the limit, the discrete graphs in Figure 15-10 would meld into the type of smooth graphs used in calculus. But the *ideas* wouldn't change. The *ideas* would be exactly as explained here. This is very similar to what you saw in Chapter 14 where you first approximated the area under a curve by adding up the areas of rectangles and then were able to compute the exact area by using the limit process to narrow the widths of the rectangles till their widths became infinitesimal.)

Well, there you have it — actual explanations of why the shortcut version of the fundamental theorem works and why finding area is differentiation in reverse. If you understand only half of what I've just written, you're way ahead of most students of calculus. The good news is that you probably won't be tested on this theoretical stuff. Now let's come back down to earth.

Finding Antiderivatives: Three Basic Techniques

I've been talking a lot about antiderivatives, but just how do you find them? In this section, I give you three easy techniques. Then in Chapter 16, I give you four advanced techniques. By the way, you *will* be tested on this stuff.

Reverse rules for antiderivatives

The easiest antiderivative rules are the ones that are the reverse of derivative rules you already know. (You can brush up on derivative rules in Chapter 10 if you need to.) These are automatic, one-step antiderivatives with the exception of the reverse power rule, which is only slightly harder.

No-brainer reverse rules

You know that the derivative of $\sin x$ is $\cos x$, so reversing that tells you that an antiderivative of $\cos x$ is $\sin x$. What could be simpler? But don't forget that all functions of the form $\sin x + C$ are antiderivatives of $\cos x$. In symbols, you write

$$\frac{d}{dx}\sin x = \cos x, \text{ and therefore}$$

$$\int \cos x \, dx = \sin x + C$$

Table 15-2 lists the reverse rules for antiderivatives.

Table 15-2	Basic Antiderivative Formulas
1) $\int 1 dx$ (or just $\int dx) = x + C$ (because the derivative of x is 1)	
2) $\int x^n dx = \frac{x^{n+1}}{n+1} + C$ $(n \neq -1)$	3) $\int \frac{dx}{x} = \ln|x| + C$ (rule 2 for $n = -1$)
4) $\int e^x dx = e^x + C$	5) $\int a^x dx = \frac{1}{\ln a} a^x + C$
6) $\int \sin x \, dx = -\cos x + C$	7) $\int \cos x \, dx = \sin x + C$
8) $\int \sec^2 x \, dx = \tan x + C$	9) $\int \csc^2 x \, dx = -\cot x + C$
10) $\int \sec x \tan x \, dx = \sec x + C$	11) $\int \csc x \cot x \, dx = -\csc x + C$
12) $\int \frac{dx}{\sqrt{a^2 - x^2}} = \arcsin \frac{x}{a} + C$	13) $\int \frac{dx}{a^2 + x^2} = \frac{1}{a} \arctan \frac{x}{a} + C$
14) $\int \frac{dx}{x\sqrt{x^2 - a^2}} = \frac{1}{a} \text{arcsec} \frac{|x|}{a} + C$	

The slightly more difficult reverse power rule

By the power rule for differentiation, you know that

$$\frac{d}{dx} x^3 = 3x^2, \text{ and therefore}$$

$$\int 3x^2 dx = x^3 + C$$

Here's the simple method for reversing the power rule. Use $y = 5x^4$ for your function. Recall that the power rule says to

1. **Bring the power in front where it will *multiply* the rest of the derivative.**

$$5x^4 \rightarrow 4 \cdot 5x^4$$

2. ***Reduce* the power by one and simplify.**

$$4 \cdot 5x^4 \rightarrow 4 \cdot 5x^3 = 20x^3$$

Thus, $y' = 20x^3$.

smile! calculus sucks, but you don't!!

To reverse this process, you reverse the order of the two steps and reverse the math within each step. Here's how that works for the above problem:

1. ***Increase* the power by one.**

The 3 becomes a 4.

$$20x^3 \rightarrow 20x^4$$

2. ***Divide* by the new power and simplify.**

$$20x^4 \rightarrow \frac{20}{4}x^4 = 5x^4$$

And thus you write $\int 20x^3 dx = 5x^4 + C$.

The reverse power rule does not work for a power of negative one. The reverse power rule works for all powers (including negative and decimal powers) except for a power of negative one. Instead of using the reverse power rule, you should just memorize that the antiderivative of x^{-1} is $\ln|x| + C$ (rule 3 in Table 15-2).

Test your antiderivatives by differentiating them. Especially when you're new to antidifferentiation, it's a good idea to test your antiderivatives by differentiating them — you can ignore the C. If you get back to your original function, you know your antiderivative is correct.

With the antiderivative you just found and the shortcut version of the fundamental theorem, you can determine the area under $20x^3$ between, say, 1 and 2:

$$\int 20x^3 dx = 5x^4 + C, \text{ thus}$$

$$\int_{1}^{2} 20x^3 dx = \left[5x^4\right]_1^2$$

$$= 5 \cdot 2^4 - 5 \cdot 1^4$$
$$= 80 - 5$$
$$= 75$$

Guessing and checking

The guess-and-check method works when the *integrand* (that's the expression after the integral symbol not counting the *dx*, and it's the thing you want to antidifferentiate) is close to a function that you know the reverse rule for. For example, say you want the antiderivative of cos(2*x*). Well, you know that the derivative of sine is cosine. Reversing that tells you that the antiderivative of cosine is sine. So you might think that the antiderivative of cos(2*x*) is sin(2*x*). That's your *guess*. Now *check* it by differentiating it to see if you get the original function, cos(2*x*):

$$\frac{d}{dx}\sin(2x)$$

$$=\cos(2x)\cdot 2 \quad \text{(sine rule and chain rule)}$$

$$=2\cos(2x)$$

This result is very close to the original function, except for that extra coefficient of 2. In other words, the answer is 2 times as much as what you want. Because you want a result that's half of this, just try an antiderivative that's half of your first guess: So your new guess is $\frac{1}{2}\sin(2x)$. Check this second guess by differentiating it, and you get the desired result.

Here's another example. What's the antiderivative of $(3x-2)^4$?

1. **Guess the antiderivative.**

   This looks sort of like a power rule problem, so try the reverse power rule. The antiderivative of x^4 is $\frac{1}{5}x^5$ by the reverse power rule, so your guess is $\frac{1}{5}(3x-2)^5$.

2. **Check your guess by differentiating it.**

   $$\frac{d}{dx}\left[\frac{1}{5}(3x-2)^5\right]$$

   $$=5\cdot\frac{1}{5}(3x-2)^4\cdot 3 \quad \text{(power rule and chain rule)}$$

   $$=3(3x-2)^4$$

3. **Tweak your first guess.**

   Your result, $3(3x-2)^4$ is three times too much, so make your second guess a *third* of your first guess — that's $\frac{1}{3}\cdot\frac{1}{5}(3x-2)^5$, or $\frac{1}{15}(3x-2)^5$.

4. Check your second guess by differentiating it.

$$\frac{d}{dx}\left[\frac{1}{15}(3x-2)^5\right]$$

$$=5\cdot\frac{1}{15}(3x-2)^4\cdot3 \quad \text{(power rule and chain rule)}$$

$$=(3x-2)^4$$

This checks. You're done. The antiderivative of $(3x-2)^4$ is $\frac{1}{15}(3x-2)^5+C$.

The two previous examples show that *guess and check* works well when the function you want to antidifferentiate has an argument like $3x$ or $3x+2$ (where x is raised to the *first* power) instead of a plain old x. (Recall that in a function like $\sqrt{5x}$, the $5x$ is called the *argument*.) In this case, all you have to do is tweak your guess by the *reciprocal* of the coefficient of x: the 3 in $3x+2$, for example (the 2 in $3x+2$ has no effect on your answer). In fact, for these easy problems, you don't really have to do any guessing and checking. You can immediately see how to tweak your guess. It becomes sort of a one-step process. If the function's argument is more complicated than $3x+2$ — like the x^2 in $\cos(x^2)$ — you have to try the next method, substitution.

The substitution method

If you look back at the examples of the guess and check method in the previous section, you can see why the first guess in each case didn't work. When you differentiate the guess, the chain rule produces an extra constant: 2 in the first example, 3 in the second. You then tweak the guesses with $\frac{1}{2}$ and $\frac{1}{3}$ to compensate for the extra constant.

Now say you want the antiderivative of $\cos(x^2)$ and you guess that it is $\sin(x^2)$. Watch what happens when you differentiate $\sin(x^2)$ to check it:

$$\frac{d}{dx}\sin(x^2)$$

$$=\cos(x^2)\cdot2x \quad \text{(sine rule and chain rule)}$$

$$=2x\cos(x^2)$$

Here the chain rule produces an extra $2x$ — because the derivative of x^2 is $2x$ — but if you try to compensate for this by attaching a $\frac{1}{2x}$ to your guess, it won't work. Try it.

So, guessing and checking doesn't work for antidifferentiating $\cos(x^2)$ — actually *no* method works for this simple-looking integrand (not all functions have antiderivatives) — but your admirable attempt at differentiation here reveals a new class of functions that you can antidifferentiate. Because the derivative of $\sin(x^2)$ is $2x\cos(x^2)$, the antiderivative of $2x\cos(x^2)$ must be $\sin(x^2)$. This function, $2x\cos(x^2)$, is the type of function you can antidifferentiate with the substitution method.

Keep your eyes peeled for the derivative of the function's argument. The substitution method works when the integrand contains a function and *the derivative of the function's argument* — in other words, when it contains that extra thing produced by the chain rule — or something just like it except for a constant. And the integrand must not contain any other extra stuff.

The derivative of e^{x^3} is $e^{x^3} \cdot 3x^2$ by the e^x rule and the chain rule. So, the antiderivative of $e^{x^3} \cdot 3x^2$ is e^{x^3}. And if you were asked to find the antiderivative of $e^{x^3} \cdot 3x^2$, you would know that the substitution method would work because this expression contains $3x^2$, which is the derivative of the argument of e^{x^3}, namely x^3.

By now, you're probably wondering why this is called the substitution method. I show you why in the step-by-step method below. But first, I want to point out that you don't always have to use the step-by-step method. Assuming you understand why the antiderivative of $e^{x^3} \cdot 3x^2$ is e^{x^3}, you may encounter problems where you can just see the antiderivative without doing any work. But whether or not you can just see the answers to problems like that one, the substitution method is a good technique to learn because, for one thing, it has many uses in calculus and other areas of mathematics, and for another, your teacher may require that you know it and use it. Okay, so here's how to find $\int 2x\cos(x^2)dx$ with substitution:

1. **Set u equal to the argument of the main function.**

   The argument of $\cos(x^2)$ is x^2, so you set u equal to x^2.

2. **Take the derivative of u with respect to x.**

   $$u = x^2 \text{ so } \frac{du}{dx} = 2x$$

3. **Solve for dx.**

   $$\frac{du}{dx} = \frac{2x}{1}$$
   $$du = 2xdx \quad \text{(cross multiplication)}$$
   $$\frac{du}{2x} = dx \quad \text{(dividing both sides by } 2x\text{)}$$

4. **Make the substitutions.**

   In $\int 2x\cos(x^2)dx$, u takes the place of x^2 and $\frac{du}{2x}$ takes the place of dx. So now you've got $\int 2x\cos u \frac{du}{2x}$. The two $2x$s cancel, giving you $\int \cos u\, du$.

5. **Antidifferentiate using the simple reverse rule.**

$$\int \cos u\, du = \sin u + C$$

6. **Substitute x^2 back in for u, coming full circle.**

   u equals x^2, so x^2 goes in for the u:

$$\int \cos u\, du = \sin(x^2) + C$$

   That's it. So $\int 2x\cos(x^2)dx = \sin(x^2) + C$.

If the original problem had been $\int 5x\cos(x^2)dx$ instead of $\int 2x\cos(x^2)dx$, you follow the same steps except that in Step 4, after making the substitution, you arrive at $\int 5x\cos u \frac{dx}{2x}$. The xs still cancel — that's the important thing — but after canceling you get $\int \frac{5}{2}\cos u\, du$, which has that extra $\frac{5}{2}$ in it. No worries. Just pull the $\frac{5}{2}$ through the $\int$ symbol, giving you $\frac{5}{2}\int \cos u\, du$. Now you finish this problem just as you did in Steps 5 and 6, except for the extra $\frac{5}{2}$:

$$\frac{5}{2}\int \cos u\, du = \frac{5}{2}(\sin u + C)$$
$$= \frac{5}{2}\sin u + \frac{5}{2}C$$
$$= \frac{5}{2}\sin(x^2) + \frac{5}{2}C$$

Because C is any old constant, $\frac{5}{2}C$ is still any old constant, so you can get rid of the $\frac{5}{2}$ in front of the C. That may seem somewhat (grossly?) unmathematical, but it's right. Thus, your final answer is $\frac{5}{2}\sin(x^2) + C$. You should check this by differentiating it.

Here are a few examples of antiderivatives you can do with the substitution method so you can learn how to spot them:

✔ $\int 4x^2\cos(x^3)dx$

   The derivative of x^3 is $3x^2$, but you don't have to pay any attention to the 3 in $3x^2$ or the 4 in the integrand. Because the integrand contains x^2 and no other extra stuff, substitution works. Try it.

✔ $\int 10\sec^2 x \cdot e^{\tan x}\,dx$

The integrand contains a function, $e^{\tan x}$, and the derivative of its argument ($\tan x$) — which is $\sec^2 x$. Because the integrand doesn't contain any other extra stuff (except for the 10, which doesn't matter), substitution works. Do it.

✔ $\int \dfrac{2}{3}\cos x\sqrt{\sin x}\,dx$

Because the integrand contains the derivative of $\sin x$, namely $\cos x$, and no other stuff except for the $\dfrac{2}{3}$, substitution works. Go for it.

You can do the three problems just listed with a method that combines substitution and guess-and-check (as long as your teacher doesn't insist that you show the six-step substitution solution). Try using this combo method to antidifferentiate the first example, $\int 4x^2\cos(x^3)\,dx$. First, you confirm that the integral fits the pattern for substitution — it does, as pointed out in the first item on the checklist. This confirmation is the only part substitution plays in the combo method. Now you finish the problem with the guess-and-check method:

1. **Make your guess.**

   The antiderivative of cosine is sine, so a good guess for the antiderivative of $4x^2\cos(x^3)$ is $\sin(x^3)$.

2. **Check your guess by differentiating it.**

$$\frac{d}{dx}\sin(x^3) = \cos(x^3)\cdot 3x^2 \quad \text{(sine rule and chain rule)}$$
$$= 3x^2\cos(x^3)$$

3. **Tweak your guess.**

   Your result from Step 2, $3x^2\cos(x^3)$ is $\dfrac{3}{4}$ of what you want, $4x^2\cos(x^3)$, so make your guess $\dfrac{4}{3}$ bigger (note that $\dfrac{4}{3}$ is the reciprocal of $\dfrac{3}{4}$). Your second guess is thus $\dfrac{4}{3}\sin(x^3)$.

4. **Check this second guess by differentiating it.**

   Oh, heck, skip this — your answer's got to work.

Finding Area with Substitution Problems

You can use the shortcut version of the fundamental theorem to calculate the area under a function that you integrate with the substitution method. You can do this in two ways. In the previous section, I use substitution, setting u equal to x^2, to find the antiderivative of $2x\cos(x^2)$:

$$\int 2x\cos(x^2)\,dx = \sin(x^2) + C$$

If you want the area under this curve from, say, 0.5 to 1, the fundamental theorem does the trick:

$$\int_{0.5}^{1} 2x\cos(x^2)\,dx = \left[\sin(x^2)\right]_{0.5}^{1}$$

$$= \sin(1^2) - \sin(0.5^2)$$
$$= \sin(1) - \sin(0.25)$$
$$\approx 0.841 - 0.247$$
$$\approx 0.594$$

Another method, which amounts to the same thing, is to change the limits of integration and do the whole problem in terms of u. Refer back to the six-step solution in the section "The substitution method." What follows is very similar, except that this time you're doing definite integration rather than indefinite integration. Again, you want the area given by $\int_{0.5}^{1} 2x\cos(x^2)\,dx$:

1. **Set u equal to x^2.**

2. **Take the derivative of u with respect to x.**

$$\frac{du}{dx} = 2x$$

3. **Solve for dx.**

$$dx = \frac{du}{2x}$$

4. **Determine the new limits of integration.**

$$u = x^2, \text{ so when } x = \frac{1}{2}, u = \frac{1}{4}$$
$$\text{and when } x = 1, u = 1$$

5. **Make the substitutions, including the new limits of integration, and cancel the two $2x$s.**

   (In this problem, only one of the limits is new because when $x = 1, u = 1$.)

$$\int_{0.5}^{1} 2x\cos(x^2)\,dx = \int_{0.25}^{1} 2x\cos u \frac{du}{2x}$$

$$= \int_{0.25}^{1} \cos u\,du$$

6. **Use the antiderivative and the fundamental theorem to get the desired area *without* making the switch back to x^2.**

$$\int_{0.25}^{1} \cos u \, du = [\sin u]_{0.25}^{1}$$

$$= \sin 1 - \sin 0.25$$

$$\approx 0.594$$

It's a case of six of one, half a dozen of another with the two methods; they require about the same amount of work. So you can take your pick — however most teachers and textbooks emphasize the second method, so you probably should learn it.

Chapter 16

Integration Techniques for Experts

. . . .

In This Chapter

▶ Breaking down integrals into parts and finding trigonometric integrals

▶ Returning to your roots with *SohCahToa*

▶ Understanding the As, Bs, and Cs of partial fractions

▶ LIATE: Lilliputians In Africa Tackle Elephants

. . . .

I figure it wouldn't hurt to give you a break from the kind of theoretical groundwork stuff that I lay on pretty thick in Chapter 15, so this chapter cuts to the chase and shows you just the nuts and bolts of several integration techniques. In Chapter 15, you saw three basic integration methods: the reverse rules, the guess-and-check method, and substitution. Now you graduate to four advanced techniques: integration by parts, trigonometric integrals, trigonometric substitution, and partial fractions. Ready?

Integration by Parts: Divide and Conquer

Integrating by parts is the integration version of the product rule for differentiation. Just take my word for it. The basic idea of integration by parts is to transform an integral you *can't* do into a simple product minus an integral you *can* do. Here's the formula:

Integration by parts formula: $\int u\,dv = uv - \int v\,du$

And here's a memory aid for it: In the first two chunks, $\int u\,dv$ and uv, the u and v are in alphabetical order. If you remember that, you can remember that the integral on the right is just like the one on the left, except the u and v are reversed.

Don't try to understand the formula yet. You'll see how it works in a minute. And don't worry about understanding the first example below until you get to the end of it. The integration by parts process may seem pretty convoluted your first time through it, so you've got to be patient. After you work through a couple examples, you'll see it's really not that bad at all.

The integration by parts box: The integration by parts formula contains four things: *u, v, du,* and *dv.* To help keep everything straight, organize your problems with a box like the one in Figure 16-1.

Figure 16-1:
The integration by parts box.

u	v
du	dv

For our first example, let's do $\int \sqrt{x}\ln(x)dx$. The integration by parts formula will convert this integral, which you can't do directly, into a simple product minus an integral you'll know how to do. First, you've got to split up the integrand into two chunks — one chunk becomes the *u* and the other the *dv* that you see on the left side of the formula. For this problem, the ln(*x*) will become your *u* chunk. Then everything else is the *dv* chunk, namely $\sqrt{x}dx$. (In the next section, I show you how to decide what goes into the *u* chunk; then, whatever is left over is automatically the *dv* chunk.) After rewriting the above integrand, you've got the following for the left side of the formula:

$$\int \underline{u\,dv}$$

$$\int \ln(x)\sqrt{x}dx$$

Now it's time to do the box thing. For each new problem, you should draw an empty four-square box, then put your *u* (ln(*x*) in this problem) in the upper-left square and your *dv* ($\sqrt{x}dx$ in this problem) in the lower-right square. See Figure 16-2.

Figure 16-2:
Filling in the box.

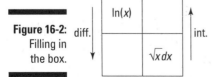

diff.   int.

ln(x)	
	$\sqrt{x}\,dx$

Next, you differentiate u to get your du, and you integrate dv to get your v. The arrows in Figure 16-2 remind you to differentiate on the left and to integrate on the right. Think of differentiation — the easier thing — as going down (like going downhill), and integration — the harder thing — as going up (like going uphill).

Now complete the box:

$$u = \ln(x) \qquad\qquad dv = \sqrt{x}\,dx$$

$$\frac{du}{dx} = \frac{1}{x} \qquad\qquad \int dv = \int \sqrt{x}\,dx$$

$$du = \frac{1}{x}\,dx \qquad\qquad v = \frac{2}{3}x^{3/2} \quad \text{(reverse power rule; note that you drop the } C\text{)}$$

Figure 16-3 shows the completed box.

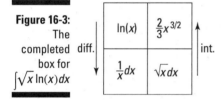

diff. | int.

You can also use the four-square box to help you remember the right side of the integration-by-parts formula: start in the upper-left square and draw (or just picture) a number 7 going straight across to the right, then down diagonally to the left. See Figure 16-4.

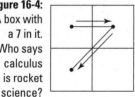

Remembering how you "draw" the 7, look back to Figure 16-3. The right side of the integration-by-parts formula tells you to do the top part of the 7, namely $\ln(x) \cdot \frac{2}{3}x^{3/2}$ minus the integral of the diagonal part of the 7, $\int \frac{2}{3}x^{3/2} \cdot \frac{1}{x}\,dx$. By the way, all of this is *much* easier to do than to explain. Try it. You'll see how this four-square-box scheme helps you learn the formula and organize these problems.

Ready to finish? Plug everything into the formula:

$$\int u\,dv = uv - \int v\,du$$

$$\int \sqrt{x}\,\ln(x)dx = \ln(x) \cdot \frac{2}{3}x^{3/2} - \int \frac{2}{3}x^{3/2} \cdot \frac{1}{x}\,dx$$

$$= \frac{2}{3}x^{3/2}\ln(x) - \frac{2}{3}\int x^{1/2}dx$$

$$= \frac{2}{3}x^{3/2}\ln(x) - \frac{2}{3}\left(\frac{2}{3}x^{3/2} + C\right) \qquad \text{(reverse power rule)}$$

$$= \frac{2}{3}x^{3/2}\ln(x) - \frac{4}{9}x^{3/2} - \frac{2}{3}C$$

$$= \frac{2}{3}x^{3/2}\ln(x) - \frac{4}{9}x^{3/2} + C, \text{ or}$$

$$= \frac{2}{3}\sqrt{x^3}\ln(x) - \frac{4}{9}\sqrt{x^3} + C$$

In the last step, you replace the $-\frac{2}{3}C$ with C because $-\frac{2}{3}$ times any old number is still just any old number.

Picking your u

Here's a great mnemonic device for how to choose your *u* chunk (again, once you've selected your *u*, everything else is automatically the *dv* chunk).

The LIATE mnemonic: Herbert E. Kasube came up with the acronym *LIATE* to help you choose your *u* (calculus nerds can check out Herb's article in the *American Mathematical Monthly* 90, 1983 issue):

L	Logarithmic	(like $\log(x)$)
I	Inverse trigonometric	(like $\arctan(x)$)
A	Algebraic	(like $5x^2 + 3$)
T	Trigonometric	(like $\cos(x)$)
E	Exponential	(like 10^x)

To pick your u chunk, go down this list in order; the first type of function on this list that appears in the integrand is the u.

Here are some helpful hints on how to remember the acronym *LIATE*. How about *Let's Integrate Another Tantalizing Example*. Or maybe you prefer *Lilliputians In Africa Tackle Elephants,* or *Lulu's Indigo And Turquoise Earrings.* The last one's not so good because it could also be *Lulu's Turquoise And Indigo Earrings* — whoops: Now you'll never remember it.

Here's an example. Integrate $\int \arctan(x)dx$. (Note, integration by parts sometimes works for integrands like this one that contain a single function.)

1. Go down the *LIATE* list and pick the *u*.

You see that there are no logarithmic functions in $\arctan(x)dx$, but there is an inverse trigonometric function, $\arctan(x)$. So that's your u. Everything else is your dv, namely, plain old dx.

2. Do the box thing.

See Figure 16-5 (and see Table 15-2 for the derivative of $\arctan(x)$).

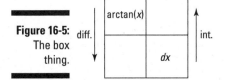

Figure 16-5: The box thing.

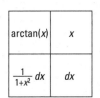

3. Plug everything into the integration by parts formula or just draw the imaginary 7 in the box on the right in Figure 16-5.

$$\int udv = uv - \int vdu$$

$$\int \arctan(x)dx = x\arctan(x) - \int x \cdot \frac{1}{1+x^2}\,dx$$

Now you can finish this problem by integrating $\int x \cdot \frac{1}{1+x^2}\,dx$ with the substitution method, setting $u = 1+x^2$. Try it (see Chapter 15 for more on the substitution method). Note that the u in $u = 1+x^2$ has nothing to do with the integration-by-parts u. Your final answer should be

$$\int \arctan(x)dx = x\arctan(x) - \frac{1}{2}\ln(1+x^2) + C.$$

Here's another one. Integrate $\int x \sin(3x)dx$:

1. **Go down the *LIATE* list and pick the *u*.**

   Going down the *LIATE* list, the first type of function you find in $x \sin(3x)dx$ is a very simple algebraic one, namely x, so that's your *u*. Everything else is your *dv*.

2. **Do the box thing.**

   See Figure 16-6.

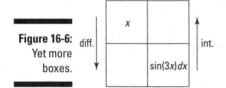

Figure 16-6: Yet more boxes.

3. **Plug everything into the integration by parts formula or draw an imaginary 7 over the box on the right in Figure 16-6.**

$$\int u\,dv = uv - \int v\,du$$

$$\int x\sin(3x)dx = -\frac{1}{3}x\cos(3x) - \int -\frac{1}{3}\cos(3x)dx$$

$$= -\frac{1}{3}x\cos(3x) + \frac{1}{3}\int\cos(3x)dx$$

You can easily integrate $\int \cos(3x)dx$ with substitution or the guess-and-check method. Go for it. Your final answer: $-\frac{1}{3}x\cos(3x) + \frac{1}{9}\sin(3x) + C$.

Integration by parts: Second time, same as the first

Sometimes you have to use the integration by parts method more than once because the first run through the method takes you only partway to the answer. Here's an example. Find $\int x^2 e^x dx$:

1. **Go down the *LIATE* list and pick the *u*.**

   $x^2 e^x dx$ contains an algebraic function, x^2, and an exponential function, e^x. (It's an exponential function because there's an x in the exponent.) The first on the *LIATE* list is x^2, so that's your *u*.

2. Do the box thing.

See Figure 16-7.

Figure 16-7:
The
boxes for
$\int x^2 e^x \, dx.$

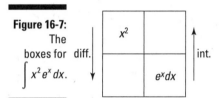

3. Use the integration by parts formula — or the "7" mnemonic.

$$\int x^2 e^x \, dx = x^2 e^x - \int e^x \cdot 2x \, dx$$
$$= x^2 e^x - 2 \int x e^x \, dx$$

You end up with another integral, $\int x^2 e^x \, dx$, that can't be done by any of the simple methods — reverse rules, guess and check, and substitution. But note that the power of x has been reduced from 2 to 1, so you've made some progress. When you use integration by parts again for $\int x e^x \, dx$, the x disappears entirely and you're done. Here goes:

4. Integrate by parts again.

I'll let you do most of this one on your own. Here's the final step:

$$\int x e^x \, dx = x e^x - \int e^x \, dx$$
$$= x e^x - e^x + C$$

5. Take the result from Step 4 and substitute it for the $\int x e^x \, dx$ in the answer from Step 3 to produce the whole enchilada.

$$\int x^2 e^x \, dx = x^2 e^x - 2(x e^x - e^x + C)$$
$$= x^2 e^x - 2x e^x + 2e^x - 2C$$
$$= x^2 e^x - 2x e^x + 2e^x + C$$

Going around in circles

Sometimes if you use integration by parts twice, you get back to where you started from — which, unlike getting lost, is *not* a waste of time. For an example of one of these odd merry-go-round type of integration by parts problems, see the free online article on the topic at www.dummies.com/extras/calculus/.

Tricky Trig Integrals

In this section (and the accompanying online article at www.dummies.com/extras/calculus/), you integrate powers of the six trigonometric functions, like $\int \sin^3(x)dx$ and $\int \sec^4(x)dx$, and products or quotients of different trig functions, like $\int \sin^2(x)\cos^3(x)dx$ and $\int \frac{\csc^2(x)}{\cot(x)} dx$. This is pretty tedious — time to order up a double espresso.

To use the following techniques, you must either have an integrand that contains just one of the six trig functions like $\int \csc^3(x)dx$ or a certain pairing of trig functions, like $\int \sin^2(x)\cos(x)dx$. If the integrand has two trig functions, the two must be one of these three pairs: sine with cosine, secant with tangent, or cosecant with cotangent. If you have an integrand containing something other than one of these three pairs, you can easily convert the problem into one of these pairs by using trig identities like $\sin(x) = \frac{1}{\csc(x)}$ and $\tan(x) = \frac{\sin(x)}{\cos(x)}$. For instance,

$$\int \sin^2(x)\sec(x)\tan(x)dx$$

$$= \int \sin^2(x) \frac{1}{\cos(x)} \cdot \frac{\sin(x)}{\cos(x)} dx$$

$$= \int \frac{\sin^3(x)}{\cos^2(x)} dx$$

After doing any needed conversions, you want to get one of the following three cases:

$$\int \sin^m(x)\cos^n(x)dx$$

$$\int \sec^m(x)\tan^n(x)dx$$

$$\int \csc^m(x)\cot^n(x)dx,$$

where either *m* or *n* (or both) is a positive integer.

The basic idea with most of the following trig integrals is to organize the integrand so that you can make a handy *u*-substitution and then integrate with the reverse power rule. You'll see what I mean in a minute.

Integrals containing sines and cosines

This section covers integrals with — can you guess? — sines and cosines.

Case 1: The power of sine is odd and positive

If the power of sine is odd and positive, lop off one sine factor and put it to the right of the rest of the expression, convert the remaining sine factors to cosines with the Pythagorean identity, and then integrate with the substitution method where $u = \cos(x)$.

The Pythagorean identity: The Pythagorean identity tells you that, for any angle x, $\sin^2(x) + \cos^2(x) = 1$. And thus $\sin^2(x) = 1 - \cos^2(x)$ and $\cos^2(x) = 1 - \sin^2(x)$.

Now integrate $\int \sin^3(x) \cos^4(x) dx$:

1. **Lop off one sine factor and move it to the right.**

$$\int \sin^3(x) \cos^4(x) dx = \int \sin^2(x) \cos^4(x) \sin(x) dx$$

2. **Convert the remaining sines to cosines using the Pythagorean identity and simplify.**

$$\int \sin^2(x) \cos^4(x) \sin(x) dx$$
$$= \int (1 - \cos^2(x)) \cos^4(x) \sin(x) dx$$
$$= \int (\cos^4(x) - \cos^6(x)) \sin(x) dx$$

3. **Integrate with the substitution method, where $u = \cos(x)$.**

$$u = \cos(x)$$
$$\frac{du}{dx} = -\sin(x)$$
$$du = -\sin(x) dx$$

Shortcut for the *u*-substitution integration method. You can save a little time in all substitution problems by just solving for *du* — as I did here — and not bothering to solve for *dx*. You then tweak the expression inside integral so that it contains the thing *du* equals and compensate for that tweaking by adding something outside the integral. In the current problem, *du* equals $-\sin(x)dx$. The integral contains a $\sin(x)dx$, so you multiply it by -1 to turn it into $-\sin(x)dx$ and then compensate for that -1 by multiplying the whole integral by -1. This is a wash because -1 times -1 equals 1. This may not sound like much of a shortcut, but it's a good time saver once you get used to it.

So tweak your integral:

$$\int (\cos^4(x) - \cos^6(x))(\sin(x)dx)$$
$$= -\int (\cos^4(x) - \cos^6(x))(-\sin(x)dx)$$

Now substitute and solve by the reverse power rule:

$$= -\int (u^4 - u^6)\,du$$
$$= -\frac{1}{5}u^5 + \frac{1}{7}u^7 + C$$
$$= -\frac{1}{5}\cos^5(x) + \frac{1}{7}\cos^7(x) + C \quad \text{or} \quad \frac{1}{7}\cos^7(x) - \frac{1}{5}\cos^5(x) + C$$

Case 2: The power of cosine is odd and positive

This problem works exactly like Case 1, except that the roles of sine and cosine are reversed. Find $\int \dfrac{\cos^3(x)}{\sqrt{\sin(x)}}\,dx$.

1. Lop off one cosine factor and move it to the right.

$$\int \frac{\cos^3(x)}{\sqrt{\sin(x)}}\,dx = \int \cos^3(x)(\sin^{-1/2}(x))dx$$
$$= \int \cos^2(x)(\sin^{-1/2}(x))\cos(x)dx$$

2. Convert the remaining cosines to sines with the Pythagorean identity and simplify.

$$\int \cos^2(x)(\sin^{-1/2}(x))\cos(x)dx$$
$$= \int (1 - \sin^2(x))(\sin^{-1/2}(x))\cos(x)dx$$
$$= \int (\sin^{-1/2}(x) - \sin^{3/2}(x))\cos(x)dx$$

3. Integrate with substitution, where $u = \sin(x)$.

$$u = \sin(x)$$

$$\frac{du}{dx} = \cos(x)$$

$$du = \cos(x)dx$$

Now substitute:

$$= \int (u^{-1/2} - u^{3/2})\, du$$

And finish integrating as in Case 1.

Case 3: The powers of both sine and cosine are even and nonnegative

Here you convert the integrand into odd powers of cosines by using the following trig identities.

Two handy trig identities:

$$\sin^2(x) = \frac{1-\cos(2x)}{2} \text{ and } \cos^2(x) = \frac{1+\cos(2x)}{2}$$

Then you finish the problem as in Case 2. Here's an example:

$$\int \sin^4(x)\cos^2(x)dx$$

$$= \int (\sin^2(x))^2 \cos^2(x)dx$$

$$= \int \left(\frac{1-\cos(2x)}{2}\right)^2 \left(\frac{1+\cos(2x)}{2}\right)dx$$

$$= \frac{1}{8}\int (1-\cos(2x)-\cos^2(2x)+\cos^3(2x))dx \quad \text{(It's just algebra!)}$$

$$= \frac{1}{8}\int 1dx - \frac{1}{8}\int \cos(2x)dx - \frac{1}{8}\int \cos^2(2x)\,dx + \frac{1}{8}\int \cos^3(2x)dx$$

The first in this string of integrals is a no-brainer; the second is a simple reverse rule with a little tweak for the 2; you do the third integral by using the $\cos^2(x)$ identity a second time; and the fourth integral is handled by following the steps in Case 2. Do it. Your final answer should be

$$\frac{1}{16}x - \frac{1}{64}\sin(4x) - \frac{1}{48}\sin^3(2x) + C$$

A veritable cake walk.

Don't forget your trig identities. If you get a sine-cosine problem that doesn't fit any of the three cases discussed above, try using a trig identity like $\sin^2(x) + \cos^2(x) = 1$ or $\cos^2(x) = \dfrac{1 + \cos(2x)}{2}$ to convert the integral into one you can handle.

For example, $\displaystyle\int \dfrac{\sin^4(x)}{\cos^2(x)}\, dx$ doesn't fit any of the three sine-cosine cases, but you can use the Pythagorean identity to convert it to $\displaystyle\int \dfrac{(1 - \cos^2(x))^2}{\cos^2(x)}\, dx = \displaystyle\int \dfrac{1 - 2\cos^2(x) + \cos^4(x)}{\cos^2(x)}\, dx$. This splits up into $\displaystyle\int \sec^2(x)\,dx - \displaystyle\int 2\,dx + \displaystyle\int \cos^2(x)\,dx$, and the rest is easy. Try it. See whether you can differentiate your result and arrive back at the original problem.

Integrals containing secants and tangents or cosecants and cotangents

The method for solving integrals containing the secant-tangent pairing or the cosecant-cotangent pairing is very similar to the method used for the sine-cosine problems. For examples, check out the online article on tricky trig integrals at www.dummies.com/extras/calculus/.

Your Worst Nightmare: Trigonometric Substitution

With the trigonometric substitution method, you can do integrals containing radicals of the following forms: $\sqrt{u^2 + a^2}$, $\sqrt{a^2 - u^2}$, and $\sqrt{u^2 - a^2}$ (as well as powers of those roots), where a is a constant and u is an expression containing x. For instance, $\sqrt{3^2 - x^2}$ is of the form $\sqrt{a^2 - u^2}$.

You're going to love this technique . . . about as much as sticking a hot poker in your eye.

Desperate times call for desperate measures. Consider pulling the fire alarm on the day your teacher is presenting this topic. With any luck, your teacher will decide that he can't afford to get behind schedule and he'll just omit this topic from your final exam.

Before I show you how trigonometric substitution works, I've got some silly mnemonic tricks to help you keep the three cases of this method straight. (Remember, with mnemonic devices, silly (and vulgar) works.) First, the three cases involve three trig functions, *tangent, sine,* and *secant*. Their initial letters, *t, s,* and *s,* are the same letters as the initial letters of the name of this technique, *trigonometric substitution*. Pretty nice, eh?

Table 16-1 shows how these three trig functions pair up with the radical forms listed in the opening paragraph.

Table 16-1	**A Totally Radical Table**
$\tan(\theta)$ $\longleftrightarrow$	$\sqrt{u^2 + a^2}$
$\sin(\theta)$ $\longleftrightarrow$	$\sqrt{a^2 - u^2}$
$\sec(\theta)$ $\longleftrightarrow$	$\sqrt{u^2 - a^2}$

To keep these pairings straight, note that the plus sign in $\sqrt{u^2 + a^2}$ looks like a little *t* for *tangent*, and that the other two forms, $\sqrt{a^2 - u^2}$ and $\sqrt{u^2 - a^2}$, contain a *subtraction* sign — *s* is for *sine* and *secant*. To memorize what sine and secant pair up with, note that $\sqrt{a^2 - u^2}$ begins with the letter *a*, and it's a *sin* to call someone an *ass*. Okay, I admit this is pretty weak. If you can come up with a better mnemonic, use it!

Ready to do some problems? I've stalled long enough.

Case 1: Tangents

Find $\int \dfrac{dx}{\sqrt{9x^2 + 4}}$. First, note that this can be rewritten as $\int \dfrac{dx}{\sqrt{(3x)^2 + 2^2}}$, so it fits the form $\sqrt{u^2 + a^2}$ where $u = 3x$ and $a = 2$; you can see that this pairs up with tangent in Table 16-1.

1. **Draw a right triangle — basically a *SohCahToa* triangle — where** $\tan(\theta)$ **equals** $\dfrac{u}{a}$**, which is** $\dfrac{3x}{2}$**.**

   Because you know that $\tan(\theta) = \dfrac{O}{A}$ (from *SohCahToa* — see Chapter 6), your triangle should have $3x$ as O, the side *opposite* the angle θ, and 2 as A, the *adjacent* side. Then, your radical, $\sqrt{(3x)^2 + 2^2}$, or $\sqrt{9x^2 + 4}$, will automatically be the correct length for the hypotenuse. It's not a bad idea to confirm this with the Pythagorean theorem, $a^2 + b^2 = c^2$. See Figure 16-8.

Figure 16-8:
A
SohCahToa
triangle for
the $\sqrt{u^2 + a^2}$
case. What
sinister
mind dreamt
up this
technique?

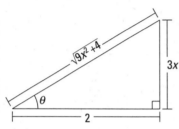

2. **Solve $\tan(\theta) = \dfrac{3x}{2}$ for x, differentiate, and solve for dx.**

$$\frac{3x}{2} = \tan(\theta)$$

$$3x = 2\tan(\theta)$$

$$x = \frac{2}{3}\tan(\theta)$$

$$\frac{dx}{d\theta} = \frac{2}{3}\sec^2(\theta)$$

$$dx = \frac{2}{3}\sec^2(\theta)d\theta$$

3. **Find which trig function is represented by the radical over the a, and then solve for the radical.**

   Look at the triangle in Figure 16-8. The radical is the *hypotenuse* and a is 2, the *adjacent* side, so $\dfrac{\sqrt{9x^2 + 4}}{2}$ is $\dfrac{H}{A}$, which equals *secant*. So $\sec(\theta) = \dfrac{\sqrt{9x^2 + 4}}{2}$, and thus $\sqrt{9x^2 + 4} = 2\sec(\theta)$.

4. **Use the results from Steps 2 and 3 to make substitutions in the original problem and then integrate.**

   From Steps 2 and 3 you have $dx = \dfrac{2}{3}\sec^2(\theta)d\theta$ and $\sqrt{9x^2 + 4} = 2\sec(\theta)$.

   Now you can finally do the integration.

$$\int \frac{dx}{\sqrt{9x^2+4}} = \int \frac{\frac{2}{3}\sec^2(\theta)d\theta}{2\sec(\theta)}$$

$$= \frac{1}{3}\int \sec(\theta)d\theta$$

$$= \frac{1}{3}\ln|\sec(\theta)+\tan(\theta)|+C \quad \text{(an integral you should memorize or just look up)}$$

5. **Substitute the x expressions from Steps 1 and 3 back in for $\sec(\theta)$ and $\tan(\theta)$. You can also get the expressions from the triangle in Figure 16-8.**

$$= \frac{1}{3}\ln\left|\frac{\sqrt{9x^2+4}}{2}+\frac{3x}{2}\right|+C$$

$$= \frac{1}{3}\ln\left|\frac{\sqrt{9x^2+4}+3x}{2}\right|+C$$

$$= \frac{1}{3}\ln\left|\sqrt{9x^2+4}+3x\right|-\frac{1}{3}\ln 2+C \quad \text{(by the log of a quotient rule, of course, and distributing the } \frac{1}{3}\text{)}$$

$$= \frac{1}{3}\ln\left|\sqrt{9x^2+4}+3x\right|+C \quad \text{(because } -\frac{1}{3}\ln 2+C \text{ is just a constant)}$$

Now tell me, when was the last time you had so much fun? Before tackling Case 2, here are a couple tips.

Step 1 is $\frac{u}{a}$. For all three cases in trigonometric substitution, Step 1 always involves drawing a triangle in which the trig function in question equals $\frac{u}{a}$:

Case 1 is $\tan(\theta)=\frac{u}{a}$.

Case 2 is $\sin(\theta)=\frac{u}{a}$.

Case 3 is $\sec(\theta)=\frac{u}{a}$.

The fact that the u goes in the numerator of this $\frac{u}{a}$ fraction should be easy to remember because u is an expression in x and something like $\frac{3x}{2}$ is somewhat simpler and more natural to see than $\frac{2}{3x}$. Just remember the x goes on top.

Step 3 is $\dfrac{\sqrt{}}{a}$. For all three cases, Step 3 always involves putting the radical over the a. The three cases are given below, but you don't need to memorize the trig functions in this list because you'll know which one you've got by just looking at the triangle — assuming you know *SohCahToa* and the reciprocal trig functions (flip back to Chapter 6 if you don't know them). I've left out what goes under the radicals because by the time you're doing Step 3, you've already got the right radical expression.

Case 1 is $\sec(\theta) = \dfrac{\sqrt{}}{a}$.

Case 2 is $\cos(\theta) = \dfrac{\sqrt{}}{a}$.

Case 3 is $\tan(\theta) = \dfrac{\sqrt{}}{a}$.

In a nutshell, just remember $\dfrac{u}{a}$ for Step 1 and $\dfrac{\sqrt{}}{a}$ for Step 3.

Case 2: Sines

Integrate $\displaystyle\int \dfrac{dx}{x^2\sqrt{16-x^2}}$, rewriting it first as $\displaystyle\int \dfrac{dx}{x^2\sqrt{4^2-x^2}}$ so that it fits the form $\sqrt{a^2-u^2}$, where $a=4$ and $u=x$.

1. **Draw a right triangle where** $\sin(\theta) = \dfrac{u}{a}$, **which is** $\dfrac{x}{4}$.

   Sine equals $\dfrac{\text{O}}{\text{H}}$, so the *opposite* side is x and the *hypotenuse* is 4. The length of the adjacent side is then automatically equal to your radical, $\sqrt{16-x^2}$. (Confirm this with the Pythagorean theorem.) See Figure 16-9.

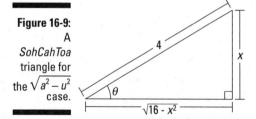

2. **Solve** $\sin(\theta) = \dfrac{x}{4}$ **for x, differentiate, and solve for dx.**

$$\frac{x}{4} = \sin(\theta)$$
$$x = 4\sin(\theta)$$
$$\frac{dx}{d\theta} = 4\cos(\theta)$$
$$dx = 4\cos(\theta)d\theta$$

3. **Find which trig function equals the radical over the a, and then solve for the radical.**

Look at the triangle in Figure 16-9. The radical, $\sqrt{16-x^2}$, over the a, 4, is $\frac{A}{H}$, which you know from *SohCahToa* equals *cosine*. That gives you

$$\cos(\theta) = \frac{\sqrt{16-x^2}}{4}$$
$$\sqrt{16-x^2} = 4\cos(\theta)$$

4. **Use the results from Steps 2 and 3 to make substitutions in the original problem and then integrate.**

Note that you have to make three substitutions here, not just two like in the first example. From Steps 2 and 3 you've got

$$x = 4\sin(\theta),\ dx = 4\cos(\theta)d\theta,\ \text{and}\ \sqrt{16-x^2} = 4\cos(\theta),\ \text{so}$$

$$\int \frac{dx}{x^2\sqrt{16-x^2}} = \int \frac{4\cos(\theta)d\theta}{(4\sin(\theta))^2 4\cos(\theta)}$$
$$= \int \frac{d\theta}{16\sin^2(\theta)}$$
$$= \frac{1}{16}\int \csc^2(\theta)d\theta$$
$$= -\frac{1}{16}\cot(\theta) + C$$

5. **The triangle shows that $\cot(\theta) = \dfrac{\sqrt{16-x^2}}{x}$. Substitute back for your final answer.**

$$= -\frac{1}{16}\cdot\frac{\sqrt{16-x^2}}{x} + C$$
$$= -\frac{\sqrt{16-x^2}}{16x} + C$$

It's a walk in the park.

Case 3: Secants

In the interest of space — and sanity — I'm going to skip this case. But you won't have any trouble with it because all the steps are basically the same as in Cases 1 and 2.

Try this one. Integrate $\int \dfrac{\sqrt{x^2-9}}{x}\,dx$. I'll get you started. In Step 1, you draw a triangle, where $\sec(\theta)=\dfrac{u}{a}$, that's $\dfrac{x}{3}$. Now take it from there. Here's the answer (no peeking if you haven't done it yet): $\sqrt{x^2-9}-3\arctan\left(\dfrac{\sqrt{x^2-9}}{3}\right)+C$, or $\sqrt{x^2-9}-3\operatorname{arcsec}\left(\dfrac{x}{3}\right)+C$, or $\sqrt{x^2-9}-3\arccos\left(\dfrac{3}{x}\right)+C$.

The As, Bs, and Cxs of Partial Fractions

Just when you thought it couldn't get any worse than trigonometric substitution, I give you the partial fractions technique.

You use the partial fractions method to integrate rational functions like $\dfrac{6x^2+3x-2}{x^3+2x^2}$. The basic idea involves "unadding" a fraction: Adding works like this: $\dfrac{1}{2}+\dfrac{1}{3}=\dfrac{5}{6}$. So, you can "unadd" $\dfrac{5}{6}$ by splitting it up into $\dfrac{1}{2}$ plus $\dfrac{1}{3}$. This is what you do with the partial fraction technique except that you do it with complicated rational functions instead of ordinary fractions.

Before using the partial fractions technique, you have to check that your integrand is a "proper" fraction — that's one where the degree of the numerator is less than the degree of the denominator. If the integrand is "improper," like $\int \dfrac{2x^3+x^2-10}{x^3-3x-2}\,dx$, you first have to do long polynomial division to transform the improper fraction into a sum of a polynomial (which sometimes will be just a number) and a proper fraction. Here's the division for this improper fraction. Basically, it works like regular long division:

$$
\begin{array}{r}
2 \\
x^3-3x-2{\overline{\smash{\big)}\,2x^3+x^2+0x-10}} \\
\underline{2x^3-6x-4} \\
x^2+6x-6
\end{array}
$$

With regular division, if you divide, say, 23 (the dividend) by 4 (the divisor), you get a quotient of 5 and a remainder of 3, which tells you that $\frac{23}{4}$ equals $5 + \frac{3}{4}$, or $5\frac{3}{4}$. The four pieces in the above polynomial division (the dividend, the divisor, the quotient, and the remainder) work the same way. The quotient is 2 and the remainder is $x^2 + 6x - 6$, thus $\frac{2x^3 + x^2 - 10}{x^3 - 3x - 2}$ equals $2 + \frac{x^2 + 6x - 6}{x^3 - 3x - 2}$. The original problem, $\int \frac{2x^3 + x^2 - 10}{x^3 - 3x - 2} dx$, therefore becomes $\int 2\,dx + \int \frac{x^2 + 6x - 6}{x^3 - 3x - 2} dx$. The first integral is just $2x + C$. You would then do the second integral with the partial fractions method. Here's how it works. First a basic example and then a more advanced one.

Case 1: The denominator contains only linear factors

Integrate $\int \frac{5}{x^2 + x - 6} dx$. This is a Case 1 problem because the factored denominator (see Step 1) contains only *linear* factors — in other words, *first* degree polynomials.

1. **Factor the denominator.**

$$\frac{5}{x^2 + x - 6} = \frac{5}{(x-2)(x+3)}$$

2. **Break up the fraction on the right into a sum of fractions, where each factor of the denominator in Step 1 becomes the denominator of a separate fraction. Then put capital-letter unknowns in the numerator of each fraction.**

$$\frac{5}{(x-2)(x+3)} = \frac{A}{(x-2)} + \frac{B}{(x+3)}$$

3. **Multiply both sides of this equation by the left side's denominator.**

This is algebra I, so you can't possibly want to see the steps, right?

$$5 = A(x+3) + B(x-2)$$

4. **Take the roots of the linear factors and plug them — one at a time — into x in the equation from Step 3, and solve for the capital-letter unknowns.**

If $x = 2$,

$5 = A(2+3) + B(2-2)$

$5 = 5A$

$A = 1$

If $x = -3$,

$5 = (-3+3) + B(-3-2)$

$5 = -5B$

$B = -1$

5. **Plug these results into the A and B in the equation from Step 2.**

$$\frac{5}{(x-2)(x+3)} = \frac{1}{(x-2)} + \frac{-1}{(x+3)}$$

6. **Split up the original integral into the partial fractions from Step 5 and you're home free.**

$$\int \frac{5}{x^2+x-6}\,dx = \int \frac{1}{(x-2)}\,dx + \int \frac{-1}{(x+3)}\,dx$$

$$= \ln|x-2| - \ln|x+3| + C$$

$$= \ln\left|\frac{x-2}{x+3}\right| + C \quad \text{(the log of a quotient rule)}$$

Case 2: The denominator contains irreducible quadratic factors

Sometimes you can't factor a denominator all the way down to linear factors because some quadratics are irreducible — like prime numbers, they can't be factored.

Check the *discriminant*. You can easily check whether a quadratic $(ax^2 + bx + c)$ is reducible or not by checking its discriminant, $b^2 - 4ac$. If the discriminant is negative, the quadratic is irreducible. If the discriminant is a perfect square like 0, 1, 4, 9, 16, 25, etc., the quadratic can be factored into factors like you're used to seeing like $(2x-5)(x+5)$. This is what happens in a Case 1 problem. The last possibility is that the discriminant equals a non-square positive number, as with the quadratic $x^2 + 10x + 1$, for example, that has a discriminant of 96. In that case, the quadratic can be factored, but you get ugly factors involving square roots. You almost certainly will not get a problem like that.

Using the partial fractions technique with irreducible quadratics is a bit different. Here's a problem: Integrate $\int \dfrac{5x^3+9x-4}{x(x-1)(x^2+4)}\,dx$.

1. **Factor the denominator.**

   It's already done! Don't say I never did anything for you. Note that x^2+4 is irreducible because its discriminant is negative.

2. **Break up the fraction into a sum of "partial fractions."**

   If you have an irreducible quadratic factor (like the x^2+4), the numerator for that partial fraction needs two capital-letter unknowns instead of just one. You write them in the form of $Px+Q$.

   $$\frac{5x^3+9x-4}{x(x-1)(x^2+4)} = \frac{A}{x} + \frac{B}{x-1} + \frac{Cx+D}{x^2+4}$$

3. **Multiply both sides of this equation by the left-side denominator.**

   $$5x^3+9x-4 = A(x-1)(x^2+4)+B(x)(x^2+4)+(Cx+D)(x)(x-1)$$

4. **Take the roots of the linear factors and plug them — one at a time — into x in the equation from Step 3, and then solve.**

If $x=0$,	If $x=1$,
$-4=-4A$	$10=5B$
$A=1$	$B=2$

   Unlike in the Case 1 example, you can't solve for all the unknowns by plugging in the roots of the linear factors, so you have more work to do.

5. **Plug into the Step 3 equation the known values of A and B and any two values for x not used in Step 4 (low numbers make the arithmetic easier) to get a system of two equations in C and D.**

   $A=1$ and $B=2$, so

If $x=-1$,	If $x=2$,
$-18=-10-10-2C+2D$	$54=8+32+4C+2D$
$2=-2C+2D$	$14=4C+2D$
$1=-C+D$	$7=2C+D$

6. **Solve the system: $1 = -C + D$ and $7 = 2C + D$.**

   You should get $C = 2$ and $D = 3$.

7. **Split up the original integral and integrate.**

   Using the values obtained in Steps 4 and 6, $A = 1$, $B = 2$, $C = 2$, and $D = 3$, and the equation from Step 2, you can split up the original integral into three pieces:

   $$\int \frac{5x^3 + 9x - 4}{x(x-1)(x^2+4)}\,dx = \int \frac{1}{x}\,dx + \int \frac{2}{x-1}\,dx + \int \frac{2x+3}{x^2+4}\,dx$$

   And with simple algebra, you can split up the third integral on the right into two pieces, resulting in the final partial fraction decomposition:

   $$\int \frac{5x^3 + 9x - 4}{x(x-1)(x^2+4)}\,dx = \int \frac{1}{x}\,dx + \int \frac{2}{x-1}\,dx + \int \frac{2x}{x^2+4}\,dx + \int \frac{3}{x^2+4}\,dx$$

   The first two integrals are easy. For the third, you use substitution with $u = x^2 + 4$ and $du = 2x\,dx$. The fourth is done with the arctangent rule, which you should memorize: $\int \frac{dx}{a^2 + x^2} = \frac{1}{a}\arctan\frac{x}{a} + C$.

   $$\int \frac{5x^3 + 9x - 4}{x(x-1)(x^2+4)}\,dx = \ln|x| + 2\ln|x-1| + \ln|x^2+4| + \frac{3}{2}\arctan\left(\frac{x}{2}\right) + C$$

   $$= \ln|x(x-1)^2(x^2+4)| + \frac{3}{2}\arctan\left(\frac{x}{2}\right) + C$$

Case 3: The denominator contains repeated linear or quadratic factors

It's likely that you won't get one of these messier problems. But if you'd like to see an example, check out the online article, "Partial Fractions Technique Where the Denominator Contains Repeated Linear or Quadratic Factors" at www.dummies.com/extras/calculus/.

Bonus: Equating coefficients of like terms

Here's another method for finding the capital-letter unknowns that you should have in your bag of tricks. Say you get the following for your Step 3 equation (this comes from a problem with two irreducible quadratic factors):

$$2x^3 + x^2 - 5x + 4 = (Ax + B)(x^2 + 1) + (Cx + D)(x^2 + 2x + 2)$$

This equation has no linear factors, so you can't plug in the roots to get the unknowns. Instead, expand the right side of the equation:

$$2x^3 + x^2 - 5x + 4 = Ax^3 + Ax + Bx^2 + B + Cx^3 + 2Cx^2 + 2Cx + Dx^2 + 2Dx + 2D$$

And collect like terms:

$$2x^3 + x^2 - 5x + 4 = (A+C)x^3 + (B+2C+D)x^2 + (A+2C+2D)x + (B+2D)$$

Then equate the coefficients of like terms from the left and right sides of the equation:

$$2 = A + C$$
$$1 = B + 2C + D$$
$$-5 = A + 2C + 2D$$
$$4 = B + 2D$$

You then solve this system of simultaneous equations to get A, B, C, and D.

How about a shortcut? You can finish the Case 2 example a couple pages back by using a shortcut version of the equating of coefficients method. Once you have the values for A and B from Step 4, you could look back at the equation in Step 3, and equate the coefficients of the x^3 term on the left and right sides of the equation. Can you see, without actually doing the expansion, that on the right you'd get $(A+B+C)x^3$? So, $5x^3 = (A+B+C)x^3$, which means that $5 = A+B+C$, and because $A = 1$ and $B = 2$ (from Step 4), C must equal 2. Then, using these values for A, B, and C, and any value of x (other than 0 or 1), you can get D. How about that for a simple shortcut?

Practice makes perfect. In a nutshell, you have three ways to find your capital-letter unknowns: 1) Plugging in the roots of the linear factors of the denominator if there are any, 2) Plugging in other values of x and solving the resulting system of equations, and 3) Equating the coefficients of like terms. With practice, you'll get good at combining these methods to find your unknowns quickly.

Chapter 17

Forget Dr. Phil: Use the Integral to Solve Problems

*A*s I say in Chapter 14, integration is basically just adding up small pieces of something to get the total for the whole thing — *really* small pieces, actually, *infinitely* small pieces. Thus, the integral

$$\int_{5\,sec.}^{20\,sec.} little\ piece\ of\ distance$$

tells you to add up all the little pieces of distance traveled during the 15-second interval from 5 to 20 seconds to get the total distance traveled during that interval.

In all problems, the little piece after the integration symbol is always an expression in x (or some other variable). For the above integral, for instance, the little piece of distance might be given by, say, $x^2 dx$, Then the definite integral

$$\int_{5}^{20} x^2 dx$$

would give you the total distance traveled. Because you're now an expert at computing integrals like the one immediately above, that's no longer the issue; your main challenge in this chapter is simply to come up with the algebraic expression for the little pieces you're adding up. But before we begin the adding-up problems, I want to cover a couple other integration topics: mean value and average value.

The Mean Value Theorem for Integrals and Average Value

The best way to understand the mean value theorem for integrals is with a diagram — look at Figure 17-1.

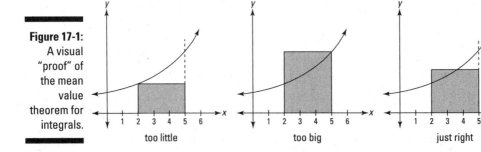

Figure 17-1: A visual "proof" of the mean value theorem for integrals.

too little   too big   just right

The graph on the left in Figure 17-1 shows a rectangle whose area is clearly *less than* the area under the curve between 2 and 5. This rectangle has a height equal to the lowest point on the curve in the interval from 2 to 5. The middle graph shows a rectangle whose height equals the highest point on the curve. Its area is clearly *greater than* the area under the curve. By now you're thinking, "Isn't there a rectangle taller than the short one and shorter than the tall one whose area is *the same as* the area under the curve?" Of course. And this rectangle obviously crosses the curve somewhere in the interval. This so-called "mean value rectangle," shown on the right, basically sums up the mean value theorem for integrals. It's really just common sense. But here's the mumbo jumbo.

The mean value theorem for integrals: If $f(x)$ is a continuous function on the closed interval $[a, b]$, then there exists a number c in the closed interval such that

$$\int_a^b f(x)dx = f(c) \cdot (b-a)$$

The theorem basically just guarantees the existence of the mean value rectangle. (Note that there can only be one mean value rectangle, but its top will sometimes cross the function more than once. Thus, there can be more than one c value that satisfies the theorem.)

The area of the mean value rectangle — which is the same as the area under the curve — equals *length* times *width,* or *base* times *height,* right? So, if you divide its area, $\int_a^b f(x)dx$ by its base, $(b-a)$, you get its height, $f(c)$. This height is the *average value* of the function over the interval in question.

Average value: The *average value* of a function $f(x)$ over a closed interval $[a, b]$ is

$$\frac{1}{b-a} \int_a^b f(x)dx$$

which is the height of the mean value rectangle.

Here's an example. What's the average speed of a car between $t=9$ seconds and $t=16$ seconds whose speed in feet per second is given by the function $f(t)=30\sqrt{t}$? The definition of average value gives you the answer in one step: the average speed is $\frac{1}{16-9} \int_9^{16} 30\sqrt{t}\, dt$. Evaluate that integral and you're done.

(That's all there is to it, so the two-step process below is somewhat superfluous. However, it shows the logic underlying the average value idea.)

1. **Determine the area under the curve between 9 and 16.**

$$\int_9^{16} 30\sqrt{t}\, dt$$

$$=30\left[\frac{2}{3}t^{3/2}\right]_9^{16}$$

$$=30\left(\frac{128}{3}-\frac{54}{3}\right)$$

$$=740$$

This area, by the way, is the total distance traveled during the period from 9 to 16 seconds, namely 740 feet. Do you see why? Consider the mean value rectangle for this problem. Its height is a speed (because the function values, or heights, are speeds) and its base is an amount of time, so its area is *speed* times *time* which equals *distance*. Alternatively, recall that the derivative of position is velocity (see Chapter 12). So, the anti-derivative of velocity — what I just did in this step — is position, and the change of position from 9 to 16 seconds gives the total distance traveled.

2. **Divide this area, total distance, by the time interval from 9 to 16, namely 7.**

$$Average\ speed = \frac{total\ distance}{total\ time} = \frac{740\ feet}{7\ seconds} \approx 105.7\ feet\ per\ second$$

The definition of average value tells you to multiply the total area by $\frac{1}{b-a}$, which in this problem is $\frac{1}{16-9}$, or $\frac{1}{7}$. But because dividing by 7 is the same as multiplying by $\frac{1}{7}$, you can divide like I do in this step. It makes more sense to think about these problems in terms of division: area equals *base* times *height*, so the height of the mean value rectangle equals its area *divided* by its base.

The MVT for integrals and for derivatives: Two peas in a pod

Remember the mean value theorem for derivatives from Chapter 11? The graph on the left in the figure shows how it works for the function $f(x) = x^3$. The basic idea is that there's a point on the curve between 0 and 2 where the slope is the same as the slope of the secant line from $(0, 0)$ to $(2, 8)$ — that's a slope of 4. When you do the math, you get $x = \frac{2\sqrt{3}}{3}$ for this point. Well, it turns out that the point guaranteed by the mean value theorem for integrals — the point where the mean value rectangle crosses the

derivative of this curve (shown on the right in the figure) — has the very same x-value. Pretty nice, eh?

If you really want to understand the intimate relationship between differentiation and integration, think long and hard about the many connections between the two graphs in the accompanying figure. This figure is a real gem, if I do say so myself. (For more on the differentiation/ integration connection, check out my other favorites, Figures 15-9 and 15-10.)

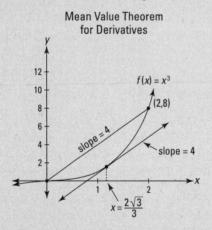

Mean Value Theorem
for Derivatives

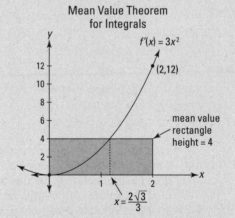

Mean Value Theorem
for Integrals

- At $x = \frac{2\sqrt{3}}{3}$ the *slope* is 4 and that's the average *slope* of f between 0 and 2.

- The least *slope* of f in the interval is 0.

- The greatest *slope* of f in the interval is 12.

- The total *rise* along f from 0 to 2 is 8.

- At $x = \frac{2\sqrt{3}}{3}$ the *height* is 4 and that's the average *height* of f' between 0 and 2.

- The least *height* of f' in the interval is 0.

- The greatest *height* of f' in the interval is 12.

- The total *area* under f' from 0 to 2 is 8.

The Area between Two Curves — Double the Fun

This is the first of several topics in this chapter where your task is to come up with an expression for a little bit of something, then add up the bits by integrating. For this first problem type, the little bit is a narrow rectangle that sits on one curve and goes up to another. Here's an example: Find the area between $y = 2 - x^2$ and $y = \frac{1}{2}x$ from $x = 0$ to $x = 1$. See Figure 17-2.

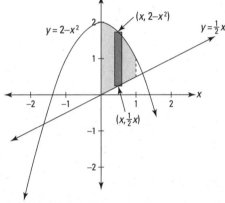

Figure 17-2: The area between $y = 2 - x^2$ and $y = \frac{1}{2}x$ from $x = 0$ to $x = 1$.

To get the height of the representative rectangle in Figure 17-2, subtract the y-coordinate of its bottom from the y-coordinate of its top — that's $(2 - x^2) - \frac{1}{2}x$. Its base is the infinitesimal dx. So, because *area* equals *height* times *base*,

$$\text{Area of representative rectangle} = ((2 - x^2) - \frac{1}{2}x)dx$$

Now you just add up the areas of all the rectangles from 0 to 1 by integrating:

$$\int_0^1 ((2 - x^2) - \frac{1}{2}x)dx$$

$$= \left[2x - \frac{1}{3}x^3 - \frac{1}{4}x^2\right]_0^1 \quad \text{(power rule for all 3 pieces)}$$

$$= \left(2 - \frac{1}{3} - \frac{1}{4}\right) - (0 - 0 - 0)$$

$$= \frac{17}{12} \text{ square units}$$

Now to make things a little more twisted, in the next problem the curves cross (see Figure 17-3). When this happens, you have to split the total shaded area into two separate regions before integrating. Try this one: Find the area between $\sqrt[3]{x}$ and x^3 from $x=0$ to $x=2$.

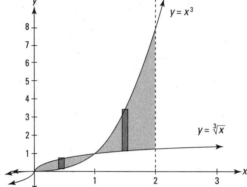

1. Determine where the curves cross.

They cross at $(1, 1)$, so you've got two separate regions: one from 0 to 1 and another from 1 to 2.

2. Figure the area of the region on the left.

For this region, $\sqrt[3]{x}$ is above x^3. So the height of a representative rectangle is $\sqrt[3]{x}-x^3$, its area is *height* times *base*, or $(\sqrt[3]{x}-x^3)dx$, and the area of the region is, therefore,

$$\int_0^1 (\sqrt[3]{x}-x^3)dx$$

$$=\left[\frac{3}{4}x^{4/3}-\frac{1}{4}x^4\right]_0^1$$

$$=\left(\frac{3}{4}-\frac{1}{4}\right)-(0-0)$$

$$=\frac{1}{2}$$

3. Figure the area of the region on the right.

In the right-side region, x^3 is above $\sqrt[3]{x}$, so the height of a rectangle is $x^3-\sqrt[3]{x}$ and thus you've got

$$\int_{1}^{2} (x^3 - \sqrt[3]{x})dx$$

$$= \left[\frac{1}{4}x^4 - \frac{3}{4}x^{4/3} \right]_{1}^{2}$$

$$= (4 - \frac{3}{2}\sqrt[3]{2}) - \left(\frac{1}{4} - \frac{3}{4} \right)$$

$$= 4.5 - 1.5\sqrt[3]{2}$$

$$\approx 2.61$$

4. Add up the areas of the two regions to get the total area.

$$0.5 + \sim 2.61 \approx 3.11 \text{ square units}$$

Height equals top minus bottom. Note that the height of a representative rectangle is always its *top* minus its *bottom,* regardless of whether these numbers are positive or negative. For instance, a rectangle that goes from 20 up to 30 has a height of $30 - 20$, or 10; a rectangle that goes from –3 up to 8 has a height of $8 - (-3)$, or 11; and a rectangle that goes from –15 up to –10 has a height of $-10 - (-15)$, or 5.

If you think about this top-minus-bottom method for figuring the height of a rectangle, you can now see — assuming you didn't already see it — why the definite integral of a function counts area below the x-axis as negative. (I mention this in Chapters 14 and 15.) For example, consider Figure 17-4.

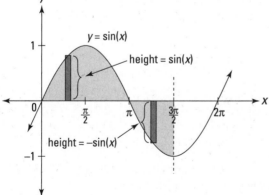

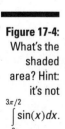

Figure 17-4:
What's the
shaded
area? Hint:
it's not
$$\int_{0}^{3\pi/2} \sin(x)dx.$$

If you want the total area of the shaded region shown in Figure 17-4, you have to divide the shaded region into two separate pieces like you did in the last problem. One piece goes from 0 to π, and the other from π to $\frac{3\pi}{2}$.

For the first piece, from 0 to π, a representative rectangle has a height equal to the function itself, $y = \sin(x)$, because its top is on the function and its bottom is at zero — and of course, anything minus zero is itself. So the area of this first piece is given by the ordinary definite integral $\int_0^\pi \sin(x)dx$.

But for the second piece from π to $\frac{3\pi}{2}$, the top of a representative rectangle is at zero — recall that the x-axis is the line $y = 0$ — and its bottom is on $y = \sin(x)$, so its height is $0 - \sin(x)$, or just $-\sin(x)$. So, to get the area of this second piece, you figure the definite integral of the *negative* of the function, $\int_\pi^{3\pi/2} (-\sin(x))dx$, which is the same as $-\int_\pi^{3\pi/2} \sin(x)dx$.

Because this *negative* integral gives you the ordinary, *positive* area of the piece below the x-axis, the *positive* definite integral $\int_\pi^{3\pi/2} \sin(x)dx$ gives a *negative* area.

That's why if you figure the definite integral $\int_0^{3\pi/2} \sin(x)dx$ over the entire span, the piece below the x-axis counts as a negative area, and the answer gives you the *net* of the area above the x-axis minus the area below the axis — rather than the total shaded area.

Finding the Volumes of Weird Solids

In geometry, you learned how to figure the volumes of simple solids like boxes, cylinders, and spheres. Integration enables you to calculate the volumes of an endless variety of much more complicated shapes.

The meat-slicer method

This metaphor is actually quite accurate. Picture a hunk of meat being cut into very thin slices on one of those deli meat slicers. That's the basic idea here. You slice up a three-dimensional shape, then add up the volumes of the slices to determine the total volume.

Here's a problem: What's the volume of the solid whose length runs along the x-axis from 0 to π and whose cross sections perpendicular to the x-axis are equilateral triangles such that the midpoints of their bases lie on the x-axis and their top vertices are on the curve $y = \sin(x)$? Is that a mouthful or what? This problem is almost harder to describe and to picture than it is to do. Take a look at this thing in Figure 17-5.

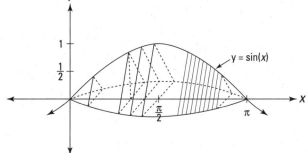

So what's the volume?

1. **Determine the area of any old cross section.**

   Each cross section is an equilateral triangle with a height of $\sin(x)$. (The height of the second triangle from the left is shown in the figure.) If you do the geometry, you'll see that the base of each triangle is $\dfrac{2\sqrt{3}}{3}$ times its height, or $\dfrac{2\sqrt{3}}{3} \cdot \sin(x)$. (Hint: Half of an equilateral triangle is a 30°-60°-90° triangle.) So, the triangle's area, given by $A = \dfrac{1}{2}(b)(h)$ is

   $$\frac{1}{2}\left(\frac{2\sqrt{3}}{3} \cdot \sin(x)\right)\sin(x), \text{ or } \frac{\sqrt{3}}{3}\sin^2(x).$$

2. **Find the volume of a representative slice.**

   The volume of a slice is just its cross-sectional area times its infinitesimal thickness, dx. So you've got the volume:

   $$\textit{Volume of representative slice} = \frac{\sqrt{3}}{3}\sin^2(x)dx$$

3. **Add up the volumes of the slices from 0 to π by integrating.**

   If the following seems a bit difficult, well, heck, you better get used to it. This is calculus after all. (Actually, it's not really that bad if you go through it patiently, step by step.)

$$\int_0^\pi \frac{\sqrt{3}}{3}\sin^2(x)dx$$

$$=\frac{\sqrt{3}}{3}\int_0^\pi \sin^2(x)dx$$

$$=\frac{\sqrt{3}}{3}\int_0^\pi \frac{1-\cos(2x)}{2}dx \qquad \text{(trig integrals with sines and cosines,}$$
$$\text{Case 3, from Chapter 16)}$$

$$=\frac{\sqrt{3}}{6}\left(\int_0^\pi 1\,dx - \int_0^\pi \cos(2x)dx\right)$$

$$=\frac{\sqrt{3}}{6}\left([x]_0^\pi - \left[\frac{\sin(2x)}{2}\right]_0^\pi\right)$$

$$=\frac{\sqrt{3}}{6}\left(\pi - 0 - \left(\frac{\sin(2\pi)}{2} - \frac{\sin(0)}{2}\right)\right)$$

$$=\frac{\sqrt{3}}{6}(\pi - 0 - (0-0))$$

$$=\frac{\pi\sqrt{3}}{6}$$

$$\approx 0.91 \text{ cubic units}$$

It's a ~~piece o' cake~~ slice o' meat.

The disk method

This technique is basically the same as the meat slicer method — actually it's a special case of the meat slicer method that you use when the cross-sections are all circles. Here's how it works. Find the volume of the solid — between $x=2$ and $x=3$ — generated by rotating the curve $y=e^x$ about the x-axis. See Figure 17-6.

1. **Determine the area of any old cross section.**

   Each cross section is a circle with radius e^x. So, its area is given by the formula for the area of a circle, $A=\pi r^2$. Plugging e^x into r gives you

   $$A=\pi(e^x)^2 = \pi e^{2x}$$

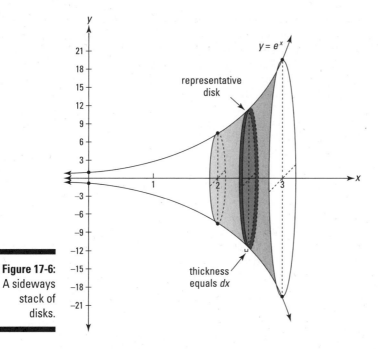

Figure 17-6:
A sideways
stack of
disks.

2. **Tack on *dx* to get the volume of an infinitely thin representative disk.**

$$Volume\ of\ disk = \overbrace{\pi e^{2x}}^{area} \cdot \overbrace{dx}^{thickness}$$

3. **Add up the volumes of the disks from 2 to 3 by integrating.**

$$Total\ volume = \int_{2}^{3} \pi e^{2x} dx$$

$$= \pi \int_{2}^{3} e^{2x} dx$$

$$= \frac{\pi}{2} \int_{2}^{3} e^{2x} 2dx \qquad \text{(The two new 2s are to tweak the integral for } \textit{the u-}\text{substitution; see next line of equation.)}$$

$$= \frac{\pi}{2} \int_{4}^{6} e^{u} du \qquad \text{(by substitution with } u = 2x \text{ and } du = 2dx; \\ \text{when } x = 2, u = 4; \text{ when } x = 3, u = 6)$$

$$= \frac{\pi}{2} \left[e^{u} \right]_{4}^{6}$$

$$= \frac{\pi}{2} \left(e^{6} - e^{4} \right)$$

$$\approx 548\ \text{cubic units}$$

TIP

A representative disk is located at no particular place. Note that Step 1 refers to "any old" cross section. I call it that because when you consider a representative disk like the one shown in Figure 17-6, you should focus on a disk that's in no place in particular. The one shown in Figure 17-6 is located at an *unknown* position on the x-axis, and its radius goes from the x-axis up to the curve $y = e^x$. Thus, its radius is the *unknown* length of e^x. If, instead, you use some special disk like the left-most disk at $x = 2$, you're more likely to make the mistake of thinking that a representative disk has some *known* radius like e^2. (This tip also applies to the meat-slicer method in the previous section and the washer method in the next section.)

The Washer Method

The only difference between the washer method and the disk method is that now each slice has a hole in its middle that you have to subtract. There's nothing to it.

Here you go. Take the area bounded by $y = x^2$ and $y = \sqrt{x}$, and generate a solid by revolving that area about the x-axis. See Figure 17-7.

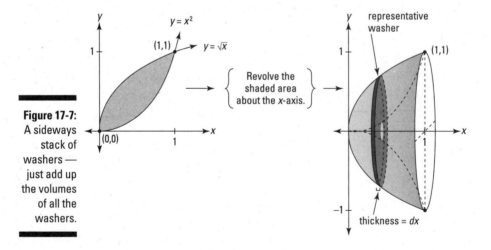

Figure 17-7: A sideways stack of washers — just add up the volumes of all the washers.

Just think: All the forces of the evolving universe and all the twists and turns of your life have brought you to *this* moment when you are finally able to calculate the volume of this solid — something for your diary. So what's the volume of this bowl-like shape?

1. **Determine where the two curves intersect.**

   It should take very little trial and error to see that $y = x^2$ and $y = \sqrt{x}$ intersect at $x = 0$ and $x = 1$ — how nice is that? So the solid in question spans the interval on the *x*-axis from 0 to 1.

2. **Figure the cross-sectional area of a thin representative washer.**

   Each slice has the shape of a washer — see Figure 17-8 — so its cross-sectional area equals the area of the entire circle minus the area of the hole.

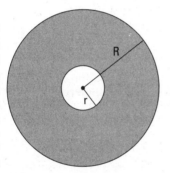

Figure 17-8:
The shaded area equals $\pi R^2 - \pi r^2$: The *whole* minus the hole — get it?

The area of the circle minus the hole is $\pi R^2 - \pi r^2$, where R is the outer radius (the big radius) and r is the hole's radius (the little radius). For this problem, the outer radius is $\sqrt{x}$ and the hole's radius is x^2, giving you

$$A = \pi \left(\sqrt{x}\right)^2 - \pi (x^2)^2$$
$$= \pi x - \pi x^4$$

3. **Multiply this area by the thickness, *dx*, to get the volume of a representative washer.**

   $$Volume = (\pi x - \pi x^4)dx$$

4. Add up the volumes of the even-thinner-than-paper-thin washers from 0 to 1 by integrating.

$$Volume = \int_0^1 (\pi x - \pi x^4)dx$$

$$= \pi \int_0^1 (x - x^4)dx$$

$$= \pi \left[\frac{1}{2}x^2 - \frac{1}{5}x^5 \right]_0^1$$

$$= \pi \left[\left(\frac{1}{2} - \frac{1}{5} \right) - (0 - 0) \right]$$

$$= \frac{3}{10}\pi$$

$$\approx 0.94 \text{ cubic units}$$

Area equals big circle minus little circle. Focus on the simple fact that the area of a washer is the area of the entire disk, πR^2, minus the area of the hole, πr^2: Thus, $Area = \pi R^2 - \pi r^2$. When you integrate, you get $\int_a^b (\pi R^2 - \pi r^2)\,dx$. If you factor out the pi, and bring it to the outside of the integral, you get $\pi \int_a^b (R^2 - r^2)\,dx$ which is the formula given in most books. But if you just learn that formula by rote, you may forget it. You're more likely to remember the formula and how to do these problems if you understand the simple big-circle-minus-little-circle idea.

The matryoshka-doll method

Another method for calculating volume (which may not be covered in your calculus course) is the cylindrical shell method. Instead of cutting up the volume in question into slices, disks, or washers, you cut it up into thin concentric cylinders. The concentric cylinders fit inside each other like those nested Russian dolls. For an example of one of these problems, check out my online article on the matryoshka doll method at www.dummies.com/go/calculus/.

Analyzing Arc Length

So far in this chapter, you've added up the areas of thin rectangles to get total area and the volumes of thin slices to get total volume. Now, you're going to add up minute lengths along a curve to get the whole length.

I could just give you the formula for *arc length* (the length along a curve), but I'd rather show you why it works and how to derive it. Lucky you. The idea is to divide a length of curve into tiny sections, figure the length of each section, and then add up all the lengths. Figure 17-9 shows how each section of a curve can be approximated by the hypotenuse of a tiny right triangle.

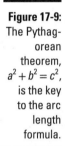

Figure 17-9:
The Pythag-
orean
theorem,
$a^2 + b^2 = c^2$,
is the key
to the arc
length
formula.

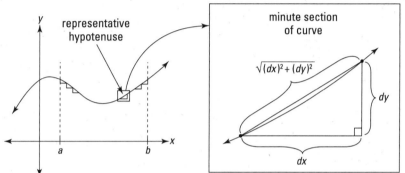

You can imagine that as you zoom in further and further, dividing the curve into more and more sections, the minute sections of the curve get straighter and straighter, and thus the perfectly straight hypotenuses become better and better approximations of the curve. That's why — when this process of adding up smaller and smaller sections is taken to the limit — you get the precise length of the curve.

So, all you have to do is add up all the hypotenuses along the curve between your start and finish points. The lengths of the legs of each infinitesimal triangle are *dx* and *dy,* and thus the length of the hypotenuse — given by the Pythagorean theorem — is

$$\sqrt{(dx)^2 + (dy)^2}$$

To add up all the hypotenuses from *a* to *b* along the curve, you just integrate:

$$\int_a^b \sqrt{(dx)^2 + (dy)^2}$$

A little tweaking and you have the formula for arc length. First, factor out a $(dx)^2$ under the square root and simplify:

$$\int_a^b \sqrt{(dx)^2 \left[1 + \frac{(dy)^2}{(dx)^2}\right]} = \int_a^b \sqrt{(dx)^2 \left[1 + \left(\frac{dy}{dx}\right)^2\right]}$$

Now you can take the square root of $(dx)^2$ — that's dx, of course — and bring it outside the radical, and, voilà, you've got the formula. . . .

Arc length formula: The arc length along a curve, $y = f(x)$, from a to b, is given by the following integral:

$$\int_a^b \sqrt{1 + \left(\frac{dy}{dx}\right)^2}\, dx$$

The expression inside this integral is simply the length of a representative hypotenuse.

Try this one: What's the length along $y = (x-1)^{3/2}$ from $x = 1$ to $x = 5$?

1. Take the derivative of your function.

$$y = (x-1)^{3/2}$$
$$\frac{dy}{dx} = \frac{3}{2}(x-1)^{1/2}$$

2. Plug this into the formula and integrate.

$$\int_a^b \sqrt{1 + \left(\frac{dy}{dx}\right)^2}\, dx$$

$$= \int_1^5 \sqrt{1 + \left(\frac{3}{2}(x-1)^{1/2}\right)^2}\, dx$$

$$= \int_1^5 \sqrt{1 + \frac{9}{4}(x-1)}\, dx$$

$$= \int_1^5 \left(\frac{9}{4}x - \frac{5}{4}\right)^{1/2} dx$$

$$= \left[\frac{4}{9} \cdot \frac{2}{3}\left(\frac{9}{4}x - \frac{5}{4}\right)^{3/2}\right]_1^5$$

(See how I got that? It's the guess-and-check integration technique with the reverse power rule. The $\frac{4}{9}$ is the tweak amount you need because of the coefficient $\frac{9}{4}$.)

$$= \left[\frac{1}{27}(9x-5)^{3/2} \right]_1^5 \quad \text{(Algebra questions are strictly prohibited!)}$$

$$= \frac{1}{27}\left(\sqrt{40}\right)^3 - \frac{1}{27}\left(\sqrt{4}\right)^3$$

$$= \frac{8}{27}\left(\left(\sqrt{10}\right)^3 - 1\right)$$

$$\approx 9.07 \text{ units}$$

Now if you ever find yourself on a road with the shape of $y=(x-1)^{3/2}$ and your odometer is broken, you can figure the exact length of your drive. Your friends will be very impressed — or very concerned.

Surfaces of Revolution — Pass the Bottle 'Round

A surface of revolution is a three-dimensional surface with circular cross sections, like a vase or a bell or a wine bottle. For these problems, you divide the surface into narrow circular bands, figure the surface area of a representative band, and then just add up the areas of all the bands to get the total surface area. Figure 17-10 shows such a shape with a representative band.

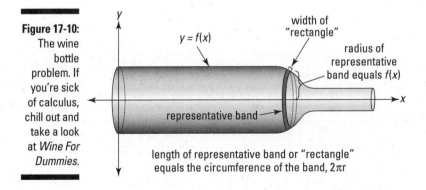

Figure 17-10: The wine bottle problem. If you're sick of calculus, chill out and take a look at *Wine For Dummies.*

y

$y = f(x)$

width of "rectangle"

radius of representative band equals $f(x)$

representative band

x

length of representative band or "rectangle" equals the circumference of the band, $2\pi r$

What's the surface area of a representative band? Well, if you cut the band and unroll it, you get sort of a long, narrow rectangle whose area, of course, is length times width. The rectangle wraps around the whole circular surface, so its length is the circumference of the circular cross section, or $2\pi r$, where r is the height of the function (for garden-variety problems anyway). The width of the rectangle or band is the same as the length of the infinitesimal hypotenuse you used in the section on arc length, namely $\sqrt{1+\left(\dfrac{dy}{dx}\right)^2}\,dx.$

Thus, the surface area of a representative band, from length times width, is

$2\pi r\sqrt{1+\left(\dfrac{dy}{dx}\right)^2}\,dx$, which brings us to the formula.

Surface of revolution formula: A surface generated by revolving a function, $y=f(x)$, about an axis has a surface area — between a and b — given by the following integral:

$$\int_a^b 2\pi r\sqrt{1+\left(\frac{dy}{dx}\right)^2}\,dx = 2\pi \int_a^b r\sqrt{1+\left(\frac{dy}{dx}\right)^2}\,dx$$

If the axis of revolution is the x-axis, r will equal $f(x)$ — as shown in Figure 17-10. If the axis of revolution is some other line, like $y=5$, it's a bit more complicated — something to look forward to.

Now try one: What's the surface area — between $x=1$ and $x=2$ — of the surface generated by revolving $y=x^3$ about the x-axis? See Figure 17-11.

Figure 17-11: A surface of revolution — this one's shaped sort of like the end of a trumpet.

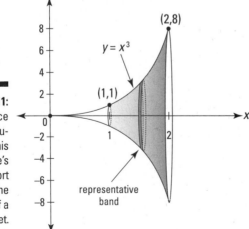

y

(2,8)

$y = x^3$

(1,1)

x

representative band

1. **Take the derivative of your function.**

$$y = x^3$$
$$\frac{dy}{dx} = 3x^2$$

Now you can finish the problem by just plugging everything into the formula, but I'll do it step by step to reinforce the idea that whenever you integrate, you write down a representative little bit of something — that's the integrand — then you add up all the little bits by integrating.

2. **Figure the surface area of a representative narrow band.**

The radius of the band is x^3, so its circumference is $2\pi x^3$ — that's the band's "length." Its width, a tiny hypotenuse, is $\sqrt{1 + \left(\dfrac{dy}{dx}\right)^2}\, dx =$

$\sqrt{1 + (3x^2)^2}\, dx$. And, thus, its area — *length* times *width* — is

$2\pi x^3 \sqrt{1 + (3x^2)^2}\, dx$.

3. **Add up the areas of all the bands from 1 to 2 by integrating.**

$$\int_1^2 2\pi x^3 \sqrt{1 + (3x^2)^2}\, dx$$

$$= 2\pi \int_1^2 x^3 \sqrt{1 + 9x^4}\, dx$$

$$= \frac{2\pi}{36} \int_1^2 36 x^3 \sqrt{1 + 9x^4}\, dx \qquad \text{(The 36 is the tweak amount for the}$$
$$\hspace{10cm} u\text{-substitution; see next line of equation.)}$$

$$= \frac{\pi}{18} \int_{10}^{145} u^{1/2}\, du \qquad \begin{array}{l}\text{(substitution with } u = 1 + 9x^4, du = 36x^3 dx; \\ \text{when } x = 1, u = 10; \text{ when } x = 2, u = 145)\end{array}$$

$$= \frac{\pi}{18} \left[\frac{2}{3} u^{3/2} \right]_{10}^{145}$$

$$= \frac{\pi}{18} \left(\frac{2}{3} \cdot 145^{3/2} - \frac{2}{3} \cdot 10^{3/2} \right)$$

$$\approx 199.5\ square\ units$$

That's a wrap.

Chapter 18

Taming the Infinite with Improper Integrals

● ●

In This Chapter

▶ The hospital rule — in case studying calculus makes you ill

▶ Meeting integrals without manners

▶ The paradox of Gabriel's horn

● ●

*I*n Chapter 17, you used down-to-earth integrals to compute some relatively ordinary things like the area between curves, the volumes of 3-D shapes, the lengths of curves, etc. I say these things are down-to-earth because all of them involved finite things — things with a beginning and an end, things (like a bowl-shaped 3-D object) that you could sort of hold in your hand. In this chapter, you enter the twilight zone of integrals that go to infinity. These are fascinating problems that often have surprising results. But first, we have to take care of L'Hôpital's rule — a handy technique to put into your calculus bag of tricks.

L'Hôpital's Rule: Calculus for the Sick

L'Hôpital's rule is a great shortcut for doing limit problems. Remember limits — from way back in Chapters 7 and 8 — like $\lim_{x \to 3} \frac{x^2 - 9}{x - 3}$? By the way, if you're wondering why I'm showing you this limit shortcut now, it's because (a) you may need it someday to solve some improper integral problems (the topic of the next section in this chapter), though we don't do such an example, and (b) you also need it for some of the infinite series problems in Chapter 19.

As with most limit problems — not counting no-brainer problems — you can't do $\lim\limits_{x \to 3} \dfrac{x^2-9}{x-3}$ with direct substitution: plugging 3 into x gives you $\dfrac{0}{0}$, which is undefined. In Chapter 8, you learned to do this problem by factoring the numerator into $(x-3)(x+3)$ and then canceling the $(x-3)$. That left you with $\lim\limits_{x \to 3}(x+3)$, which equals 6.

Now watch how easy it is to take the limit with L'Hôpital's rule. Simply take the derivative of the numerator and the derivative of the denominator. Don't use the quotient rule; just take the derivatives of the numerator and denominator separately. The derivative of x^2-9 is $2x$ and the derivative of $x-3$ is 1. L'Hôpital's rule lets you replace the numerator and denominator by their derivatives like this:

$$\lim_{x \to 3} \frac{x^2-9}{x-3} = \lim_{x \to 3} \frac{2x}{1}$$

The new limit is a no-brainer: $\lim\limits_{x \to 3} \dfrac{2x}{1} = \dfrac{2 \cdot 3}{1} = 6$

That's all there is to it. L'Hôpital's rule transforms a limit you can't do with direct substitution into one you can do with substitution. That's what makes it such a great shortcut.

Here's the mumbo jumbo.

L'Hôpital's rule: Let f and g be differentiable functions. If the limit of $\dfrac{f(x)}{g(x)}$ as x approaches c produces $\dfrac{0}{0}$ or $\dfrac{\pm\infty}{\pm\infty}$ when you substitute the value of c into x, then

$$\lim_{x \to c} \frac{f(x)}{g(x)} = \lim_{x \to c} \frac{f'(x)}{g'(x)}$$

Note that c can be a number or $\pm\infty$. And note that in the $\pm\infty$ over $\pm\infty$ case, both infinities can be of the same sign or one can be positive and the other negative.

Here's an example involving $\dfrac{\infty}{\infty}$: What's $\lim\limits_{x \to \infty} \dfrac{\ln(x)}{x}$? Direct substitution gives you $\dfrac{\infty}{\infty}$, so you can use L'Hôpital's rule. The derivative of $\ln(x)$ is $\dfrac{1}{x}$, and the derivative of x is 1, so

$$\lim_{x \to \infty} \frac{\ln(x)}{x} = \lim_{x \to \infty} \frac{\frac{1}{x}}{1} = \frac{\frac{1}{\infty}}{1} = \frac{0}{1} = 0$$

Try another one: Evaluate $\lim\limits_{x \to 0} \dfrac{e^{3x}-1}{x}$. Substitution gives you $\dfrac{0}{0}$ so L'Hôpital's rule applies. The derivative of $e^{3x}-1$ is $3e^{3x}$ and the derivative of x is 1, thus

$$\lim_{x \to 0} \frac{e^{3x}-1}{x} = \lim_{x \to 0} \frac{3e^{3x}}{1} = \frac{3 \cdot 1}{1} = 3$$

You must have zero over zero or infinity over infinity. The mumbo jumbo says that to use L'Hôpital's rule, substitution must produce either $\frac{0}{0}$ or $\frac{\pm\infty}{\pm\infty}$. You must get one of these acceptable "indeterminate" forms in order to apply the shortcut. Don't forget to check this.

Getting unacceptable forms into shape

If substitution produces one of the unacceptable forms, $\pm\infty\cdot 0$ or $\infty-\infty$, you first have to tweak the problem to get an acceptable form before using L'Hôpital's rule.

For instance, find $\lim\limits_{x\to\infty}\left(e^{-x}\sqrt{x}\right)$. Substituting infinity into x gives you $0\cdot\infty$ so you've got to tweak it:

$$\lim_{x\to\infty}\left(e^{-x}\sqrt{x}\right)=\lim_{x\to\infty}\left(\frac{\sqrt{x}}{e^x}\right)$$

Now you've got the $\frac{\infty}{\infty}$ case, so you're all set to use L'Hôpital's rule. The derivative of $\sqrt{x}$ is $\dfrac{1}{2\sqrt{x}}$, and the derivative of e^x is e^x, so

$$\lim_{x\to\infty}\left(\frac{\sqrt{x}}{e^x}\right)=\lim_{x\to\infty}\left(\frac{\frac{1}{2\sqrt{x}}}{e^x}\right)=\frac{\frac{1}{2\sqrt{\infty}}}{e^\infty}=\frac{\frac{1}{\infty}}{\infty}=\frac{0}{\infty}=0$$

Here's another problem: What's $\lim\limits_{x\to 0^+}\left(\dfrac{1}{1-\cos x}-\dfrac{1}{x}\right)$? (Recall from Chapter 7 that $\lim\limits_{x\to 0^+}$ means that x approaches 0 from the right only; this is a *one-sided* limit.) First, substitute zero into x (actually, since x is approaching zero from the right, you must imagine plugging a tiny positive number into x, or you can sort of think of it as plugging a "positive" zero into x). Substitution gives you $\left(\dfrac{1}{1-0.999999...}-\dfrac{1}{0^+}\right)$, which results in $\infty-\infty$, one of the unacceptable forms. So tweak the limit expression with some algebra:

$$\lim_{x\to 0^+}\left(\frac{1}{1-\cos x}-\frac{1}{x}\right)=\lim_{x\to 0^+}\left(\frac{x}{x(1-\cos x)}-\frac{1-\cos x}{x(1-\cos x)}\right)$$

$$=\lim_{x\to 0^+}\left(\frac{x-1+\cos x}{x(1-\cos x)}\right)$$

Now substitution gives you $\frac{0}{0}$, so you can finish with L'Hôpital's rule:

$$\lim_{x\to0^+}\left(\frac{x-1+\cos x}{x(1-\cos x)}\right)=\lim_{x\to0^+}\left(\frac{1-\sin x}{1(1-\cos x)+x(\sin x)}\right)$$

$$=\frac{1-0^+}{1\cdot0^++0^+\cdot0^+}$$

$$=\frac{1}{0^+}$$

$$=+\infty$$

That's it.

Three more unacceptable forms

When substitution of the arrow-number into the limit expression produces one of the unacceptable forms $1^{\pm\infty}$, 0^0, or ∞^0, you use the following logarithm trick to turn it into an acceptable form. Here's how it works. Let's find $\lim_{x\to0^+}(\sin x)^x$.

Substitution gives you $(\sin 0)^0$, which equals 0^0, so you do the following:

1. Set the limit equal to y.

$$y=\lim_{x\to0^+}(\sin x)^x$$

2. Take the log of both sides.

$$\ln(y)=\ln\left(\lim_{x\to0^+}(\sin x)^x\right)$$

$$=\lim_{x\to0^+}\left(\ln(\sin x)^x\right)\quad\text{(Take my word for it.)}$$

$$=\lim_{x\to0^+}(x\ln(\sin x))\quad\begin{array}{l}\text{(Better review the log rules in}\\ \text{Chapter 4 if you don't get this.)}\end{array}$$

3. This limit is a $0\cdot(-\infty)$ case, so tweak it.

$$=\lim_{x\to0^+}\left(\frac{\ln(\sin x)}{\frac{1}{x}}\right)$$

4. Now you've got a $\frac{-\infty}{\infty}$ case, so you can use L'Hôpital's rule.

The derivative of $\ln(\sin x)$ is $\frac{1}{\sin x}\cdot\cos x$, or $\cot x$, and the derivative of $\frac{1}{x}$ is $-\frac{1}{x^2}$, so

$$\lim_{x \to 0^+} \left(\frac{\ln(\sin x)}{\frac{1}{x}} \right) = \lim_{x \to 0^+} \left(\frac{\cot x}{-\frac{1}{x^2}} \right)$$

$$= \lim_{x \to 0^+} \left(\frac{-x^2}{\tan x} \right)$$

5. **This is a $\frac{0}{0}$ case, so use L'Hôpital's rule again.**

$$= \lim_{x \to 0^+} \left(\frac{-2x}{\sec^2 x} \right)$$
$$= \frac{0}{1}$$
$$= 0$$

Hold your horses! This is *not* the answer.

6. **Solve for y.**

Do you see that the answer of 0 in Step 5 is the answer to the equation from way back in Step 2: $\ln(y) = \ln\left(\lim_{x \to 0^+} (\sin x)^x \right)$? So, the 0 in Step 5 tells you that $\ln(y) = 0$. Now solve for y:

$$\ln(y) = 0$$
$$y = 1$$

Because you set your limit equal to y in Step 1, this, finally, is your answer:

$$\lim_{x \to 0^+} (\sin x)^x = 1$$

Ordinary math doesn't work with infinity (or zero to the zero power). Don't make the mistake of thinking that you can use ordinary arithmetic or the laws of exponents when dealing with any of the acceptable or unacceptable indeterminate forms. It might look like $\infty - \infty$ should equal zero, for example, but it doesn't. By the same token, $0 \cdot \infty \neq 0$, $\frac{0}{0} \neq 1$, $\frac{\infty}{\infty} \neq 1$, $0^0 \neq 1$, $\infty^0 \neq 1$, and $1^\infty \neq 1$.

Improper Integrals: Just Look at the Way That Integral Is Holding Its Fork!

Definite integrals are *improper* when they go infinitely far up, down, right, or left. They go up or down infinitely far in problems like $\int_2^4 \frac{1}{x-3} dx$ that have one or more vertical asymptotes. They go infinitely far to the right or left in problems like

$$\int_5^\infty \frac{1}{x^2} dx \text{ or } \int_{-\infty}^\infty \frac{1}{x^4+1} dx, \text{ where one or both of the limits of integration are infinite.}$$

(There are a couple other weird types of improper integrals, but they're rare — don't worry about them.) It would seem to make sense to just use the term *infinite* instead of *improper* to describe these integrals, except for the remarkable fact that many of these "infinite" integrals give you a *finite* area. More about this in a minute.

You solve both types of improper integrals by turning them into limit problems. Take a look at some examples.

Improper integrals with vertical asymptotes

There are two cases to consider here: problems where there's a vertical asymptote at one of the edges of the area in question and problems where there's a vertical asymptote somewhere in the middle of the area.

A vertical asymptote at one of the limits of integration

What's the area under $y = \frac{1}{x^2}$ from 0 to 1? This function is undefined at $x = 0$, and it has a vertical asymptote there. So you've got to turn the definite integral into a limit where c approaches the x-value of the asymptote:

$$\int_0^1 \frac{1}{x^2} dx = \lim_{c \to 0^+} \int_c^1 \frac{1}{x^2} dx \quad \text{(The area in question is to the right of zero, so } c \text{ approaches zero from the right.)}$$

$$= \lim_{c \to 0^+} \left[-\frac{1}{x}\right]_c^1 \quad \text{(reverse power rule)}$$

$$= \lim_{c \to 0^+} \left((-1) - \left(-\frac{1}{c}\right)\right)$$

$$= -1 - (-\infty)$$

$$= -1 + \infty$$

$$= \infty$$

This area is infinite, which probably doesn't surprise you because the curve goes up to infinity. But hold on to your hat — the next function also goes up to infinity at $x = 0$, but its area is finite!

Find the area under $y = \dfrac{1}{\sqrt[3]{x}}$ from 0 to 1. This function is also undefined at $x = 0$, so the process is the same as in the previous example:

$$\int_0^1 \frac{1}{\sqrt[3]{x}}\, dx = \lim_{c \to 0^+} \int_c^1 \frac{1}{\sqrt[3]{x}}\, dx$$

$$= \lim_{c \to 0^+} \left[\frac{3}{2} x^{2/3} \right]_c^1 \quad \text{(reverse power rule)}$$

$$= \lim_{c \to 0^+} \left(\frac{3}{2} - \frac{3}{2} c^{2/3} \right)$$

$$= \frac{3}{2} - 0$$

$$= \frac{3}{2}$$

Convergence and divergence: You say that an improper integral converges if the limit exists — that is, if the limit equals a finite number like in the second example. Otherwise, an improper integral is said to *diverge* — like in the first example. When an improper integral diverges, the area in question (or part of it) usually (but not always) equals ∞ or $-\infty$.

A vertical asymptote between the limits of integration

If the undefined point of the integrand is somewhere in between the limits of integration, you split the integral in two — at the undefined point — then turn each integral into a limit and go from there. Evaluate $\displaystyle\int_{-1}^8 \frac{1}{\sqrt[3]{x}}\, dx$. This integrand is undefined at $x = 0$.

1. **Split the integral in two at the undefined point.**

$$\int_{-1}^8 \frac{1}{\sqrt[3]{x}}\, dx = \int_{-1}^0 \frac{1}{\sqrt[3]{x}}\, dx + \int_0^8 \frac{1}{\sqrt[3]{x}}\, dx$$

2. **Turn each integral into a limit and evaluate.**

   For the $\displaystyle\int_{-1}^0$ integral, the area is to the left of zero, so c approaches zero from the left. For the $\displaystyle\int_0^8$ integral, the area is to the right of zero, so c approaches zero from the right.

$$= \lim_{c \to 0^-} \int_{-1}^{c} \frac{1}{\sqrt[3]{x}}\, dx + \lim_{c \to 0^+} \int_{c}^{8} \frac{1}{\sqrt[3]{x}}\, dx$$

$$= \lim_{c \to 0^-} \left[\frac{3}{2} x^{2/3} \right]_{-1}^{c} + \lim_{c \to 0^+} \left[\frac{3}{2} x^{2/3} \right]_{c}^{8}$$

$$= \lim_{c \to 0^-} \left(\frac{3}{2} c^{2/3} - \frac{3}{2} \right) + \lim_{c \to 0^+} \left(6 - \frac{3}{2} c^{2/3} \right)$$

$$= -\frac{3}{2} + 6$$

$$= 4.5$$

Keep your eyes peeled for *x*-values where an integrand is undefined. If you fail to notice that an integrand is undefined at an *x*-value between the limits of integration, and you integrate the ordinary way, you may get the wrong answer. The above problem, $\int_{-1}^{8} \frac{1}{\sqrt[3]{x}}\, dx$ (undefined at $x=0$), happens to work out correctly if you do it the ordinary way. However, if you do $\int_{-1}^{1} \frac{1}{x^2}\, dx$ (also undefined at $x=0$) the ordinary way, not only do you get the wrong answer, you get the totally absurd answer of *negative* 2, despite the fact that the area in question is above the *x*-axis and is therefore a positive area. The moral: *Don't risk it.*

If a part diverges, the whole diverges. If either part of the split up integral diverges, the original integral diverges. You can't get, say, $-\infty$ for one part and ∞ for the other part and add them up to get zero.

Improper integrals with one or two infinite limits of integration

You do these improper integrals by turning them into limits where c approaches infinity or negative infinity. Two examples: $\int_{1}^{\infty} \frac{1}{x^2}\, dx$ and $\int_{1}^{\infty} \frac{1}{x}\, dx$.

$$\int_{1}^{\infty} \frac{1}{x^2}\, dx = \lim_{c \to \infty} \int_{1}^{c} \frac{1}{x^2}\, dx$$

$$= \lim_{c \to \infty} \left[-\frac{1}{x} \right]_{1}^{c}$$

$$= \lim_{c \to \infty} \left(-\frac{1}{c} - \left(-\frac{1}{1} \right) \right)$$

$$= 0 - (-1)$$

$$= 1$$

So this improper integral *converges.*

In the next integral, the denominator is smaller, x instead of x^2, and thus the fraction is *bigger,* so you'd expect $\int_1^\infty \frac{1}{x}\, dx$ to be bigger than $\int_1^\infty \frac{1}{x^2}\, dx$, which it is. But it's not just bigger, it's *way* bigger:

$$\int_1^\infty \frac{1}{x}\, dx = \lim_{c \to \infty} \int_1^c \frac{1}{x}\, dx$$
$$= \lim_{c \to \infty} \left[\ln x\right]_1^c$$
$$= \lim_{c \to \infty} (\ln c - \ln 1)$$
$$= \infty - 0$$
$$= \infty$$

This improper integral *diverges.*

Figure 18-1 shows these two functions. The area under $\frac{1}{x^2}$ from 1 to ∞ is exactly the same as the area of the 1-by-1 square to its left: 1 square unit. The area under $\frac{1}{x}$ from 1 to ∞ is *much, much* bigger — actually, it's infinitely bigger than a square large enough to enclose the Milky Way Galaxy. Their shapes are quite similar, but their areas couldn't be more different.

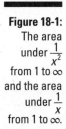

Figure 18-1:
The area under $\frac{1}{x^2}$ from 1 to ∞ and the area under $\frac{1}{x}$ from 1 to ∞.

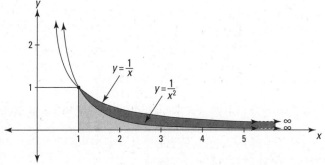

By the way, these two functions make another appearance in Chapter 19 on infinite series. Deciding whether an infinite series converges or diverges — a distinction quite similar to the difference between these two functions — is one of the main topics in Chapter 19.

When both of the limits of integration are infinite, you split the integral in two and turn each part into a limit. Splitting up the integral at $x = 0$ is convenient because zero's an easy number to deal with, but you can split it up anywhere you like. Zero may also seem like a good choice because it looks like it's in the middle between $-\infty$ and ∞. But that's an illusion because there is no middle between $-\infty$ and ∞, or you could say that any point on the x-axis is the middle.

Here's an example: $\displaystyle\int_{-\infty}^{\infty} \frac{1}{x^2+1}\,dx$

1. Split the integral in two.

$$\int_{-\infty}^{\infty} \frac{1}{x^2+1}\,dx = \int_{-\infty}^{0} \frac{1}{x^2+1}\,dx + \int_{0}^{\infty} \frac{1}{x^2+1}\,dx$$

2. Turn each part into a limit.

$$= \lim_{c\to-\infty} \int_{c}^{0} \frac{1}{x^2+1}\,dx + \lim_{c\to\infty} \int_{0}^{c} \frac{1}{x^2+1}\,dx$$

3. Evaluate each part and add up the results.

$$= \lim_{c\to-\infty} \left[\tan^{-1}x\right]_c^0 + \lim_{c\to\infty} \left[\tan^{-1}x\right]_0^c$$

$$= \lim_{c\to-\infty} (\tan^{-1}0 - \tan^{-1}c) + \lim_{c\to\infty}(\tan^{-1}c - \tan^{-1}0)$$

$$= \left(0 - \left(-\frac{\pi}{2}\right)\right) + \left(\frac{\pi}{2} - 0\right)$$

$$= \pi$$

Why don't you do this problem again, splitting up the integral somewhere other than at $x = 0$, to confirm that you get the same result.

If either "half" integral diverges, the whole, original integral diverges.

Blowing Gabriel's horn

This horn problem may blow your mind.

Gabriel's horn is the solid generated by revolving about the x-axis the unbounded region between $y = \dfrac{1}{x}$ and the x-axis (for $x \geq 1$). See Figure 18-2. Playing this instrument poses several not-insignificant challenges: 1) It has no end for you to put in your mouth; 2) Even if it did, it would take you till the end of time to reach the end; 3) Even if you could reach the end and put it in your mouth, you couldn't force any air through it because the hole is infinitely small; 4) Even if you could blow the horn, it'd be kind of pointless because it would take an infinite amount of time for the sound to come out. There are additional difficulties — infinite weight, doesn't fit in universe, and so on — but I suspect you get the picture.

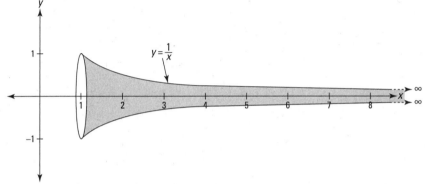

Figure 18-2:
Gabriel's
horn.

Believe it or not, Gabriel's horn has a finite volume, but an *infinite* surface area! You use the disk method to figure its volume (see Chapter 17). Recall that the volume of each representative disk is $\pi r^2 dx$. For this problem, the radius is $\frac{1}{x}$, so the little bit of volume is $\pi\left(\frac{1}{x}\right)^2 dx$. You find the total volume by adding up the little bits from 1 to ∞:

$$Volume = \int_1^\infty \pi\left(\frac{1}{x}\right)^2 dx$$

$$= \pi \int_1^\infty \frac{1}{x^2} dx$$

In the section on improper integrals, we calculated that $\int_1^\infty \frac{1}{x^2} dx = 1$, so the volume is $\pi \cdot 1$, or just π.

To determine the surface area, you first need the function's derivative (the method for calculating surface area is covered in the "Surfaces of Revolution" section in Chapter 17):

$$y = \frac{1}{x}$$

$$\frac{dy}{dx} = -\frac{1}{x^2}$$

Now plug everything into the surface area formula:

$$Surface\ Area = 2\pi \int_{1}^{\infty} \frac{1}{x} \sqrt{1 + \left(-\frac{1}{x^2}\right)^2}\ dx$$

$$= 2\pi \int_{1}^{\infty} \frac{1}{x} \sqrt{1 + \frac{1}{x^4}}\ dx$$

In the previous section, we determined that $\int_{1}^{\infty} \frac{1}{x}\ dx = \infty$, and because $\frac{1}{x}\sqrt{1 + \frac{1}{x^4}}$ is always greater than $\frac{1}{x}$ in the interval $[1, \infty)$, $\int_{1}^{\infty} \frac{1}{x}\sqrt{1 + \frac{1}{x^4}}\ dx$ must also equal ∞.

Finally, 2π times ∞ is still ∞, of course, so the surface area is infinite.

Bonus question for those with a philosophical bent: Assuming Gabriel is omnipotent, could he overcome the above-mentioned difficulties and blow this horn? *Hint:* All the calculus in the world won't help you with this one.

Chapter 19

Infinite Series

● ●

In This Chapter

▶ Segueing from sequences into series

▶ An infinite series — the rain delays just wouldn't end

▶ Getting musical with the harmonic series

▶ Taking a close look at telescoping series

▶ Rooting for the root test

▶ Testing for convergence

▶ Analyzing alternating series

● ●

A s with just about every topic in calculus, the subject of this chapter involves the idea of infinity — specifically, series that continue to infinity. An infinite series is the sum of an endless list of numbers like $\frac{1}{2} + \frac{1}{3} + \frac{1}{4} + \frac{1}{5} + \dots$. Because the list is unending, it's not surprising that such a sum can be infinite. What's remarkable is that some infinite series add up to a *finite* number. This chapter covers ten tests for deciding whether the sum of a series is finite or infinite.

What you do in this chapter is quite fantastic when you think about it. Consider the series $0.1 + 0.01 + 0.001 + 0.0001 + \dots$. If you go out far enough, you'll find a number that has so many zeros to the right of the decimal point that even if each zero were as small as a proton, there wouldn't be enough room in the entire universe just to write it down! As vast as our universe is, anything in it — say the number of elementary particles — is a proverbial drop in the bucket next to the things you look at in this chapter. Actually, not even a drop in the bucket, because next to infinity, any finite thing amounts to *nothing*. You've probably heard Carl Sagan get emotional about the "billions and billions" of stars in our galaxy. "Billions and billions" — *pffffftt.*

Sequences and Series: What They're All About

Here's a *sequence*: $\frac{1}{2}, \frac{1}{4}, \frac{1}{8}, \frac{1}{16}, \dots$. Change the commas to addition signs and you've got a *series*: $\frac{1}{2} + \frac{1}{4} + \frac{1}{8} + \frac{1}{16} + \dots$. Pretty simple, eh? Investigating series is what this chapter is all about, but I need to briefly discuss sequences to lay the groundwork for series.

Stringing sequences

A *sequence* is simply a list of numbers. An *infinite sequence* is an *unending* list of numbers. That's the only kind we're interested in here, and whenever the term *sequence* (or *series*) is used alone in this chapter, it means an infinite sequence (or infinite series).

Here's the general form for an infinite sequence:

$$a_1, a_2, a_3, a_4, \dots, a_n, \dots$$

where n runs from 1 (usually) to infinity (sometimes n starts at zero or another number). The fourth *term* of this sequence, for example, is a_4 (read "a sub 4"); the nth term is a_n (read "a sub n"). The thing we care about is what happens to a sequence infinitely far out to the right, or as mathematicians say, "in the limit." A shorthand notation for this sequence is $\{a_n\}$.

A few paragraphs back, I mentioned the following sequence. It's defined by the formula $a_n = \frac{1}{2^n}$:

$$\frac{1}{2}, \frac{1}{4}, \frac{1}{8}, \frac{1}{16}, \dots, \frac{1}{2^n}, \dots$$

What happens to this sequence as n approaches infinity should be pretty easy to see. Each term gets smaller and smaller, right? And if you go out far enough, you can find a term as close to zero as you want, right? So,

$$\lim_{n \to \infty} a_n = \lim_{n \to \infty} \frac{1}{2^n} = \frac{1}{2^\infty} = \frac{1}{\infty} = 0$$

Recall from Chapters 7 and 8 how to interpret this limit: As n approaches infinity (but never gets there), a_n gets closer and closer to zero (but never gets there).

Convergence and divergence of sequences

Because the limit of the previous sequence is a *finite* number, you say that the sequence *converges*.

Convergence and divergence: For any sequence $\{a_n\}$, if $\lim\limits_{n\to\infty} a_n = L$, where L is a real number, then the sequence *converges* to L. Otherwise, the sequence is said to *diverge*.

Sequences that converge sort of settle down to some particular number — plus or minus some miniscule amount — after you go out to the right far enough. Sequences that diverge never settle down. Instead, diverging sequences might . . .

  ✔ Increase forever, in which case $\lim\limits_{n\to\infty} a_n = \infty$. Such a sequence is said to "blow up." A sequence can also have a limit of negative infinity.

  ✔ Oscillate (go up and down) like the sequence 1, −1, 1, −1, 1, −1, 1, −1 ...

  ✔ Exhibit no pattern at all — this is rare.

Sequences and functions go hand in hand

The sequence $\left\{\dfrac{1}{2^n}\right\} = \dfrac{1}{2}, \dfrac{1}{4}, \dfrac{1}{8}, \dfrac{1}{16}, \dots, \dfrac{1}{2^n}, \dots$ can be thought of as an infinite set of discrete points (*discrete* is a fancy math word for *separate*) along the continuous function $f(x) = \dfrac{1}{2^x}$. Figure 19-1 shows the curve $f(x) = \dfrac{1}{2^x}$ and the points on the curve that make up the sequence.

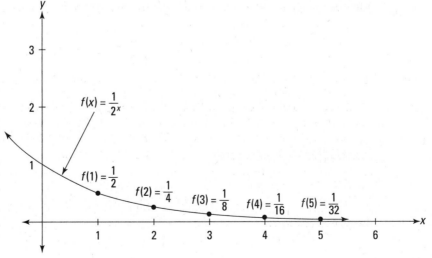

Figure 19-1: The points on the curve $f(x) = \dfrac{1}{2^x}$ make up the sequence $\left\{\dfrac{1}{2^n}\right\}$.

The sequence is made up of the outputs (the y-values) of the function where the inputs (the x-values) are positive integers $(1, 2, 3, 4, \ldots)$.

A sequence and the related function go hand in hand. If the limit of the function as x approaches infinity is some finite number, L, then the limit of the sequence is also L, and thus, the sequence converges to L. Also, the graph of such a convergent function/sequence pair has a horizontal asymptote at $y = L$; the graph in Figure 19-1 has an asymptote with the equation $y = 0$.

Determining limits with L'Hôpital's rule

Remember L'Hôpital's rule from Chapter 18? You're going to use it now to find limits of sequences. Does the sequence $a_n = \frac{n^2}{2^n}$ converge or diverge? By plugging in 1, then 2, then 3, and so on into $\frac{n^2}{2^n}$, you generate the first few terms of the sequence:

$$\frac{1}{2}, 1, \frac{9}{8}, 1, \frac{25}{32}, \frac{36}{64}, \frac{49}{128}, \frac{64}{256}, \ldots$$

What do you think? After going up for a couple terms, the sequence goes down and it appears that it'll keep going down — looks like it will converge to zero. L'Hôpital's rule proves it. You use the rule to determine the limit of the function $f(x) = \frac{x^2}{2^x}$, which goes hand in hand with the sequence $\frac{n^2}{2^n}$.

Take two separate derivatives. To use L'Hôpital's rule, take the derivative of the numerator and the derivative of the denominator separately; you do not use the quotient rule.

For this problem, you have to use L'Hôpital's rule twice:

$$\lim_{x \to \infty} \frac{x^2}{2^x} = \lim_{x \to \infty} \frac{2x}{2^x \ln 2} = \lim_{x \to \infty} \frac{2}{2^x \ln 2 \ln 2} = \frac{2}{\infty} = 0$$

Because the limit of the function is 0, so is the limit of the sequence, and thus the sequence $\frac{n^2}{2^n}$ converges to zero.

Summing series

An *infinite series* (or just *series* for short) is simply the adding up of the infinite number of terms of a sequence. Here's the sequence from the previous section again, $a_n = \frac{1}{2^n}$:

$$\frac{1}{2}, \frac{1}{4}, \frac{1}{8}, \frac{1}{16}, \ldots$$

And here's the *series* associated with this *sequence:*

$$\frac{1}{2}+\frac{1}{4}+\frac{1}{8}+\frac{1}{16}+...$$

You can use fancy summation notation to write this sum in a more compact form:

$$\sum_{n=1}^{\infty}\frac{1}{2^n}$$

The summation symbol tells you to plug 1 in for n, then 2, then 3, and so on, and then to add up all the terms (more on summation notation in Chapter 14). Nitpickers may point out that you can't actually add up an infinite number of terms. Okay, so here's the fine print for the nitpickers. An infinite sum is technically a limit. In other words,

$$\sum_{n=1}^{\infty}\frac{1}{2^n}=\lim_{b\to\infty}\sum_{n=1}^{b}\frac{1}{2^n}$$

To find an infinite sum, you take a limit — just like you do for improper (infinite) integrals (see Chapter 18). From here on, though, I just write infinite sums like $\sum_{n=1}^{\infty}\frac{1}{2^n}$ and dispense with the limit mumbo jumbo.

Partial sums

Continuing with the same series, take a look at how the sum grows by listing the "sum" of one term (kind of like the sound of one hand clapping), the sum of two terms, three terms, four, and so on:

$$S_1=\frac{1}{2}$$

$$S_2=\frac{1}{2}+\frac{1}{4}=\frac{3}{4}$$

$$S_3=\frac{1}{2}+\frac{1}{4}+\frac{1}{8}=\frac{7}{8}$$

$$S_4=\frac{1}{2}+\frac{1}{4}+\frac{1}{8}+\frac{1}{16}=\frac{15}{16}$$

.
.
.

$$S_n=\frac{1}{2}+\frac{1}{4}+\frac{1}{8}+\frac{1}{16}+\frac{1}{32}+\frac{1}{64}+.....+\frac{1}{2^n}=\frac{2^n-1}{2^n}$$

Each of these sums is called a *partial sum* of the series.

Partial sum: The nth *partial sum*, S_n, of an infinite series is the sum of the first n terms of the series.

The convergence or divergence of a series — the main event

If you now list the preceding partial sums, you have the following *sequence* of partial sums:

$$\frac{1}{2}, \frac{3}{4}, \frac{7}{8}, \frac{15}{16}, \dots$$

The main point of this chapter is figuring out whether such a sequence of partial sums *converges* — homes in on a finite number — or *diverges*. If the sequence of partial sums converges, you say that the series converges; otherwise, the sequence of partial sums diverges and you say that the series diverges. The rest of this chapter is devoted to the many techniques used in making this determination.

By the way, if you're getting a bit confused by the terms *sequence* and *series* and the connection between them, you're not alone. Keeping the ideas straight can be tricky. For starters, note that there are two sequences associated with every series. With the series $\frac{1}{2} + \frac{1}{4} + \frac{1}{8} + \frac{1}{16} + \dots$, for example, you have the underlying sequence, $\frac{1}{2}, \frac{1}{4}, \frac{1}{8}, \frac{1}{16}, \dots$, and also the sequence of partial sums, $\frac{1}{2}, \frac{3}{4}, \frac{7}{8}, \frac{15}{16}, \dots$. It's not a bad idea to try to keep these things straight, but all you really need to worry about is whether the *series* adds up to some finite number or not. If it does, it *converges*; if not, it *diverges*. The reason for getting into the somewhat confusing notion of a *sequence* of partial sums is that the definitions of convergence and divergence are based on the behavior of sequences, not series. But — I hope it goes without saying — ideas are more important than terminology, and again, the important *idea* you need to focus on is whether or not a series sums up to a finite number.

What about the previous series? Does it converge or diverge? It shouldn't take too much imagination to see the following pattern:

$$S_1 = \frac{1}{2} = 1 - \frac{1}{2}$$

$$S_2 = \frac{3}{4} = 1 - \frac{1}{4}$$

$$S_3 = \frac{7}{8} = 1 - \frac{1}{8}$$

$$S_4 = \frac{15}{16} = 1 - \frac{1}{16}$$

$$\vdots$$

$$S_n = 1 - \frac{1}{2^n}$$

Finding the limit of this sequence of partial sums is a no brainer:

$$\lim_{n \to \infty} S_n = \lim_{n \to \infty} \left(1 - \frac{1}{2^n}\right) = 1 - \frac{1}{\infty} = 1 - 0 = 1$$

So, this series converges to 1. In symbols,

$$\sum_{n=1}^{\infty} \frac{1}{2^n} = \frac{1}{2} + \frac{1}{4} + \frac{1}{8} + \frac{1}{16} + \dots\dots = 1$$

By the way, this may remind you of that paradox about walking toward a wall, where your first step is halfway to the wall, your second step is half of the remaining distance, your third step is half the remaining distance, and so on. Will you ever get to the wall? Answer: It depends. More about that later.

Convergence or Divergence? That Is the Question

This section contains nine ways of determining whether a series converges or diverges. (Then, in the next section on alternating series, we look at a final tenth test for convergence/divergence.) Note that all of the series we investigate in this section are made up of positive terms.

A no-brainer divergence test: The nth term test

If the individual terms of a series (in other words, the terms of the series' underlying sequence) do not converge to zero, then the series must diverge. This is the nth term test for divergence.

The nth term test: If $\lim_{n \to \infty} a_n \neq 0$, then $\sum a_n$ diverges. (I presume you figured out that with this naked summation symbol, n runs from 1 to infinity.)

(**Note:** The nth term test not only works for ordinary positive series like the ones in this section, it also works for series with positive and negative terms. More about this at the end of this chapter in the "Alternating Series" section.)

If you think about it, the nth term test is just common sense. When a series converges, the sum is homing in on a certain number. The only way this can happen is when the numbers being added are getting infinitesimally small — like in the series I've been talking about: $\frac{1}{2} + \frac{1}{4} + \frac{1}{8} + \frac{1}{16} + \dots$.

Imagine, instead, that the terms of a series are converging, say, to 1, like in the series $\frac{1}{2}+\frac{2}{3}+\frac{3}{4}+\frac{4}{5}+\frac{5}{6}+\dots$, generated by the formula $a_n = \frac{n}{n+1}$. In that case, when you add up the terms, you are adding numbers extremely close to 1 over and over and over forever — and this must add up to infinity. So, in order for a series to converge, the terms of the series must converge to zero. But make sure you understand what this nth term test does *not* say.

When the terms of a series converge to zero, that does *not* guarantee that the series converges. In hifalutin' logicianese — the fact that the terms of a series converge to zero is a *necessary* but *not sufficient* condition for concluding that the series converges to a finite sum.

Because this test is often very easy to apply, it should be one of the first things you check when trying to determine whether a series converges or diverges. For example, if you're asked to determine whether $\sum\limits_{n=1}^{\infty}\left(1+\frac{1}{n}\right)^n$ converges or diverges, note that every term of this series is a number greater than 1 being raised to a positive power. This always results in a number greater than 1, and thus, the terms of this series do not converge to zero, and the series must therefore diverge.

Three basic series and their convergence/divergence tests

In this subsection, we look at geometric series, p-series, and telescoping series. Geometric series and p-series are relatively simple but important series that, in addition to being interesting in their own right, can be used as benchmarks when determining the convergence or divergence of more complicated series. Telescoping series don't come up much, but many calculus texts describe them, so who am I to buck tradition?

Geometric series

A geometric series is a series of the form

$$a+ar+ar^2+ar^3+ar^4+\dots=\sum\limits_{n=0}^{\infty}ar^n$$

The first term, a, is called the *leading term*. Each term after the first equals the preceding term multiplied by r, which is called the *ratio*.

For example, if a is 5 and r is 3, you get

$$5+5\cdot3+5\cdot3^2+5\cdot3^3+\dots=5+15+45+135+\dots$$

You just multiply each term by 3 to get the next term. By the way, the 3 in this example is called the *ratio* because the ratio of any term *divided* by its preceding term equals 3, but I think it makes a lot more sense to think of the 3 as your *multiplier*.

If a is 100 and r is 0.1, you get

$$100 + 100 \cdot 0.1 + 100 \cdot 0.1^2 + 100 \cdot 0.1^3 + 100 \cdot 0.1^4 + \ldots$$
$$= 100 + 10 + 1 + 0.1 + 0.01 + \ldots$$

If that rings a bell, you've got a good memory. It's the series for the Achilles versus the tortoise paradox (go way back to Chapter 2).

And if a is $\frac{1}{2}$ and r is also $\frac{1}{2}$, you get the series I've been talking so much about:

$$\frac{1}{2} + \frac{1}{4} + \frac{1}{8} + \frac{1}{16} + \ldots$$

The convergence/divergence rule for geometric series is a snap.

Geometric series rule: If $0 < |r| < 1$, the geometric series $\sum\limits_{n=0}^{\infty} ar^n$ converges to $\frac{a}{1-r}$. If $|r| \geq 1$, the series diverges. (Note that this rule works when $-1 < r < 0$, in which case you get an *alternating series*; more about that at the end of this chapter.)

In the first example, $a = 5$ and $r = 3$, so the series diverges. In the second example, a is 100 and r is 0.1, so the series converges to $\frac{100}{1-0.1} = \frac{100}{0.9} = 111\frac{1}{9}$. That's the answer to the Achilles versus the tortoise problem: Achilles passes the tortoise after running $111\frac{1}{9}$ meters. And in the third example, $a = \frac{1}{2}$ and $r = \frac{1}{2}$, so the series converges to $\frac{1/2}{1-1/2} = 1$. This is how far you walk if you start 1 yard from the wall, then step half way to the wall, then half of the remaining distance, and so on and so on. You take an infinite number of steps, but travel a mere yard. And how long will it take you to get to the wall? Well, if you keep up a constant speed and don't pause between steps (which, of course, is impossible), you'll get there in the same amount of time it would take you to walk any old yard. If you do pause between steps, even for a billionth of a second, you'll *never* get to the wall.

p-series

A *p*-series is of the form

$$\sum_{n=1}^{\infty} \frac{1}{n^p} = \frac{1}{1^p} + \frac{1}{2^p} + \frac{1}{3^p} + \frac{1}{4^p} + \ldots$$

(where p is a *positive* power). The p-series for $p=1$ is called the *harmonic* series. Here it is:

$$\frac{1}{1} + \frac{1}{2} + \frac{1}{3} + \frac{1}{4} + \frac{1}{5} + \frac{1}{6} + \ldots$$

Although this grows *very* slowly — after 10,000 terms, the sum is only about 9.79! — the harmonic series in fact diverges to infinity.

By the way, this is called a *harmonic* series because the numbers in the series have something to do with the way a musical string like a guitar string vibrates — don't ask. For history buffs, in the 6th century B.C., Pythagoras investigated the harmonic series and its connection to the notes of the lyre.

Here's the convergence/divergence rule for p-series:

p-series rule: The p-series $\sum_{n=1}^{\infty} \frac{1}{n^p}$ converges if $p > 1$ and diverges if $p \leq 1$.

As you can see from this rule, the harmonic series forms the convergence/divergence borderline for p-series. Any p-series with terms *larger* than the terms of the harmonic series *diverges,* and any p-series with terms *smaller* than the terms of the harmonic series *converges.*

The p-series for $p=2$ is another common series:

$$1 + \frac{1}{2^2} + \frac{1}{3^2} + \frac{1}{4^2} + \frac{1}{5^2} + \frac{1}{6^2} + \ldots$$
$$= 1 + \frac{1}{4} + \frac{1}{9} + \frac{1}{16} + \frac{1}{25} + \frac{1}{36} + \ldots$$

The p-series rule tells you that this series converges. Note, however, that the p-series rule can't tell you what number this series converges to. (Contrast that to the geometric series rule, which can answer both questions.) By other means — beyond the scope of this book — it can be shown that this sum converges to $\frac{\pi^2}{6}$.

Telescoping series

You don't see many telescoping series, but the telescoping series rule is a good one to keep in your bag of tricks — you never know when it might come in handy. Consider the following series:

$$\sum_{n=1}^{\infty} \frac{1}{n(n+1)} = \frac{1}{2} + \frac{1}{6} + \frac{1}{12} + \frac{1}{20} + \frac{1}{30} + \ldots$$

To see that this is a telescoping series, you have to use the partial fractions technique from Chapter 16 — sorry to have to bring that up again — to rewrite $\dfrac{1}{n(n+1)}$ as $\dfrac{1}{n} - \dfrac{1}{n+1}$. Now you've got

$$\sum_{n=1}^{\infty} \left(\frac{1}{n} - \frac{1}{n+1} \right) = \left(1 - \frac{1}{2} \right) + \left(\frac{1}{2} - \frac{1}{3} \right) + \left(\frac{1}{3} - \frac{1}{4} \right) + \left(\frac{1}{4} - \frac{1}{5} \right) + \dots + \left(\frac{1}{n} - \frac{1}{n+1} \right)$$

Do you see how all these terms will now collapse, or *telescope?* The $\frac{1}{2}$s cancel, the $\frac{1}{3}$s cancel, the $\frac{1}{4}$s cancel, and so on. All that's left is the first term, 1, (actually, it's only half a term) and the "last" half-term, $-\dfrac{1}{n+1}$. So the sum of the first n terms is simply $1 - \dfrac{1}{n+1}$. In the limit, as n approaches infinity, $\dfrac{1}{n+1}$ converges to zero, and thus the sum converges to $1 - 0$, or 1.

Each term in a telescoping series can be written as the difference of two half-terms — call them h-terms. The telescoping series can then be written as

$$(h_1 - h_2) + (h_2 - h_3) + (h_3 - h_4) + (h_4 - h_5) + \dots + (h_n - h_{n+1}) + \dots$$

I bet you're dying for another rule, so here's the next one.

Telescoping series rule: A telescoping series of the above form converges if h_{n+1} converges to a finite number. In that case, the series converges to $h_1 - \lim_{n\to\infty} h_{n+1}$. If h_{n+1} diverges, the series diverges.

Note that this rule, like the rule for geometric series, lets you determine what number a convergent telescoping series converges to. These are the only two rules I cover where you can do this. The other eight rules for determining convergence or divergence don't allow you to determine what a convergent series converges to. But hey, you know what they say, "two out of ten ain't bad."

Three comparison tests for convergence/divergence

Say you're trying to figure out whether a series converges or diverges, but it doesn't fit any of the tests you know. No worries. You find a benchmark series that you know converges or diverges and then compare your new series to the known benchmark. For the next three tests, if the benchmark converges, your series converges; and if the benchmark diverges, your series diverges.

The direct comparison test

This is a simple, common-sense rule. If you've got a series that's *smaller* than a *convergent* benchmark series, then your series must also converge. And if your series is *larger* than a *divergent* benchmark series, then your series must also diverge. Here's the mumbo jumbo.

Direct comparison test: Let $0 \le a_n \le b_n$ for all n.

If $\sum\limits_{n=1}^{\infty} b_n$ converges, then $\sum\limits_{n=1}^{\infty} a_n$ converges.

If $\sum\limits_{n=1}^{\infty} a_n$ diverges, then $\sum\limits_{n=1}^{\infty} b_n$ diverges.

How about an example? Determine whether $\sum\limits_{n=1}^{\infty} \dfrac{1}{5+3^n}$ converges or diverges. Piece o' cake. This series resembles $\sum\limits_{n=1}^{\infty} \dfrac{1}{3^n}$, which is a geometric series with r equal to $\dfrac{1}{3}$. (Note that you can rewrite this in the standard geometric series form as $\sum\limits_{n=0}^{\infty} \dfrac{1}{3}\left(\dfrac{1}{3}\right)^n$.) Because $0 < |r| < 1$, this series converges. And because $\dfrac{1}{5+3^n}$ is *less* than $\dfrac{1}{3^n}$ for all values of n, $\sum\limits_{n=1}^{\infty} \dfrac{1}{5+3^n}$ must also converge.

Here's another one: Does $\sum\limits_{n=1}^{\infty} \dfrac{\ln n}{n}$ converge or diverge? This series resembles $\sum\limits_{n=1}^{\infty} \dfrac{1}{n}$, the harmonic p-series that is known to diverge. Because $\dfrac{\ln n}{n}$ is *greater* than $\dfrac{1}{n}$ for all values of $n \ge 3$, then $\sum\limits_{n=1}^{\infty} \dfrac{\ln n}{n}$ must also diverge. By the way, if you're wondering why I'm allowed to consider only the terms where $n \ge 3$, here's why:

Feel free to ignore initial terms. For any of the convergence/divergence tests, you can disregard *any* number of terms at the beginning of a series. And if you're comparing two series, you can ignore any number of terms from the beginning of either or both of the series — and you can ignore a different number of terms in each of the two series.

This utter disregard of innocent beginning terms is allowed because the first, say, 10 or 1,000 or 1,000,000 terms of a series always sum to a finite number and thus never have any effect on whether the series converges or diverges. Note, however, that disregarding a number of terms *would* affect the total that a convergent series converges to.

(Are you wondering why this disregard of beginning terms doesn't violate the direct comparison test's requirement that $0 \le a_n \le b_n$ for *all* n? Everything's copacetic because you can lop off any number of terms at the beginning of each series and let the counter, n, start at 1 anywhere in each series. Thus the "first" terms a_1 and b_1 can actually be located anywhere along each series. See what I mean?)

Fore! (That was a joke.) The direct comparison test tells you *nothing* if the series you're investigating is *greater* than a known *convergent* series or *less* than a known *divergent* series.

For example, say you want to determine whether $\sum_{n=1}^{\infty} \dfrac{1}{10+\sqrt{n}}$ converges. This series resembles $\sum_{n=1}^{\infty} \dfrac{1}{\sqrt{n}}$, which is a *p*-series with *p* equal to $\dfrac{1}{2}$. The *p*-series test says that this series diverges, but that doesn't help you because your series is *less* than this known divergent benchmark.

Instead, you should compare your series to the divergent harmonic series, $\sum_{n=1}^{\infty} \dfrac{1}{n}$. Your series, $\dfrac{1}{10+\sqrt{n}}$, is greater than $\dfrac{1}{n}$ for all $n \geq 14$ (it takes a little work to show this; give it a try). Because your series is *greater* than the *divergent* harmonic series, your series must also diverge.

The limit comparison test

The idea behind this test is that if you take a known convergent series and multiply each of its terms by some number, then that new series also converges. And it doesn't matter whether that multiplier is, say, 100, or 10,000, or $\dfrac{1}{10,000}$ because any number, big or small, times the finite sum of the original series is still a finite number. The same thing goes for a divergent series multiplied by any number. That new series also diverges because any number, big or small, times infinity is still infinity. This is oversimplified — it's only in the limit that one series is sort of a multiple of the other — but it conveys the basic principle.

You can discover whether such a connection exists between two series by looking at the ratio of the *n*th terms of the two series as *n* approaches infinity. Here's the test.

Limit comparison test: For two series, $\sum a_n$ and $\sum b_n$, if $a_n > 0$, $b_n > 0$, and $\lim_{n \to \infty} \left(\dfrac{a_n}{b_n} \right) = L$, where L is *finite* and *positive,* then either both series converge or both diverge.

Use this test when your series goes the wrong way. This is a good test to use when you can't use the direct comparison test for your series because it goes the wrong way — in other words, your series is *larger* than a known *convergent* series or *smaller* than a known *divergent* series.

Here's an example: Does $\sum\limits_{n=2}^{\infty} \dfrac{1}{n^2 - \ln n}$ converge or diverge? This series resembles the convergent *p*-series, $\dfrac{1}{n^2}$, so that's your benchmark. But you can't use the direct comparison test because the terms of your series are *greater* than $\dfrac{1}{n^2}$. Instead, you use the limit comparison test.

Take the limit of the ratio of the *n*th terms of the two series. It doesn't matter which series you put in the numerator and which in the denominator, but putting the known, benchmark series in the denominator makes it a little easier to do these problems and grasp the results.

$$\lim_{n \to \infty} \frac{\dfrac{1}{n^2 - \ln n}}{\dfrac{1}{n^2}}$$

$$= \lim_{n \to \infty} \frac{n^2}{n^2 - \ln n}$$

$$= \lim_{n \to \infty} \frac{2n}{2n - \dfrac{1}{n}} \qquad \text{(L'Hôpital's rule)}$$

$$= \lim_{n \to \infty} \frac{2}{2 + \dfrac{1}{n^2}} \qquad \text{(L'Hôpital's rule again)}$$

$$= \frac{2}{2 + \dfrac{1}{\infty}}$$

$$= \frac{2}{2 + 0}$$

$$= 1$$

Because the limit is finite and positive and because the benchmark series converges, your series must also converge. Thus, $\sum\limits_{n=2}^{\infty} \dfrac{1}{n^2 - \ln n}$ converges.

Use this test for rational functions. The limit comparison test is a good one for series where the general term is a *rational* function — in other words, where the general term is a quotient of two polynomials.

For example, determine the convergence or divergence of $\sum\limits_{n=1}^{\infty} \dfrac{5n^2 - n + 1}{n^3 + 4n + 3}$.

1. Determine the benchmark series.

Take the highest power of *n* in the numerator and the denominator — ignoring any coefficients and all other terms — then simplify. Like this:

$$\frac{5n^2 - n + 1}{n^3 + 4n + 3} \rightarrow \frac{n^2}{n^3} = \frac{1}{n}$$

That's the benchmark series, $\dfrac{1}{n}$, the *divergent* harmonic series.

2. **Take the limit of the ratio of the nth terms of the two series.**

$$\lim_{n\to\infty} \frac{\dfrac{5n^2 - n + 1}{n^3 + 4n + 3}}{\dfrac{1}{n}}$$

$$= \lim_{n\to\infty} \frac{5n^3 - n^2 + n}{n^3 + 4n + 3}$$

$$= \lim_{n\to\infty} \frac{5 - \dfrac{1}{n} + \dfrac{1}{n^2}}{1 + \dfrac{4}{n^2} + \dfrac{3}{n^3}} \quad \text{(dividing numerator and denominator by } n^3\text{)}$$

$$= \frac{5 - \dfrac{1}{\infty} + \dfrac{1}{\infty}}{1 + \dfrac{4}{\infty} + \dfrac{3}{\infty}}$$

$$= \frac{5 - 0 + 0}{1 + 0 + 0}$$

$$= 5$$

3. **Because the limit from Step 2 is finite and positive and because the benchmark series diverges, your series must also diverge.**

Thus, $\displaystyle\sum_{n=1}^{\infty} \frac{5n^2 - n + 1}{n^3 + 4n + 3}$ diverges.

Okay, so I'm a rebel. The limit comparison test is always stated as it appears at the beginning of this section, but I want to point out — recklessly ignoring the noble tradition of calculus textbook authors — that in a sense it's incomplete. The limit, *L*, doesn't have to be finite and positive for the test to work. First, if the benchmark series is convergent, and you put it in the denominator of the limit, and the limit is *zero*, then your series must also converge. Note that if the limit is infinity, you can't conclude anything. And second, if the benchmark series is divergent, and you put it in the denominator, and the limit is *infinity*, then your series must also diverge. If the limit is zero, you can't conclude anything.

The integral comparison test

The third benchmark test involves comparing the series you're investigating to its companion improper integral (see Chapter 18 for more on improper integrals). If the integral converges, your series converges; and if the integral diverges, so does your series. By the way, to the best of my knowledge, no one else calls this the integral comparison test — but they should because that's the way it works.

Here's an example. Determine the convergence or divergence of $\displaystyle\sum_{n=2}^{\infty} \frac{1}{n \ln n}$.

The direct comparison test doesn't work because this series is *smaller* than the *divergent* harmonic series, $\frac{1}{n}$. Trying the limit comparison test is the next natural choice, but it doesn't work either — try it. But if you notice that the series is an expression you know how to integrate, you're home free (you did notice that, right?). Just compute the companion improper integral with the same limits of integration as the index numbers of the summation — like this:

$$\int_{2}^{\infty} \frac{1}{x \ln x}\, dx$$

$$= \lim_{b \to \infty} \int_{2}^{b} \frac{1}{x \ln x}\, dx$$

$$= \lim_{b \to \infty} \int_{\ln 2}^{\ln b} \frac{1}{u}\, du \qquad \text{(substitution with } u = \ln x \text{ and } du = \frac{1}{x}\, dx;$$
$$\text{when } x = 2, u = \ln 2, \text{ and } x = b, u = \ln b)$$

$$= \lim_{b \to \infty} \left[\ln u\right]_{\ln 2}^{\ln b}$$

$$= \lim_{b \to \infty} (\ln(\ln b) - \ln(\ln 2))$$

$$= (\ln(\ln \infty) - \ln(\ln 2)$$

$$= \infty - \ln(\ln 2)$$

$$= \infty$$

Because the integral diverges, your series diverges.

After you've determined the convergence or divergence of a series with the integral comparison test, you can then use that series as a benchmark for investigating other series with the direct comparison or limit comparison tests.

For instance, the integral test just told you that $\sum_{n=2}^{\infty} \frac{1}{n \ln n}$ diverges. Now you can use this series to investigate $\sum_{n=3}^{\infty} \frac{1}{n \ln n - \sqrt{n}}$ with the direct comparison test. Do you see why? Or you can investigate, say, $\sum_{n=1}^{\infty} \frac{1}{n \ln n + \sqrt{n}}$ with the limit comparison test. Try it.

Don't forget the integral test. The integral comparison test is fairly easy to use, so don't neglect to ask yourself whether you can integrate the series expression or something close to it. If you can, it's a BINGO.

By the way, in Chapter 18, you saw the following two improper integrals:

$\int_1^\infty \frac{1}{x}dx$, which diverges and $\int_1^\infty \frac{1}{x^2}dx$, which converges. Look back at Figure 18-1.

Now that you know the integral comparison test, you can appreciate the connection between those integrals and their companion *p*-series: the divergent

harmonic series, $\sum_{n=1}^\infty \frac{1}{n}$, and the convergent *p*-series, $\sum_{n=1}^\infty \frac{1}{n^2}$.

Here's the mumbo jumbo for the integral comparison test. Note the fine print.

Integral comparison test: If $f(x)$ is positive, continuous, and decreasing for all

$x \geq 1$ and if $a_n = f(n)$, then $\sum_{n=1}^\infty a_n$ and $\int_1^\infty f(x)dx$ either both converge or both diverge.

The two "R" tests: Ratios and roots

Unlike the three benchmark tests from the previous section, the ratio and root tests don't compare a new series to a known benchmark. They work by looking only at the nature of the series you're trying to figure out. They form a cohesive pair because the results of both tests tell you the same thing. If the result is less than 1, the series converges; if it's more than 1, the series diverges; and if it's exactly 1, you learn nothing and must try a different test. (As presented here, the ratio and root tests are used for series of *positive* terms. In other books, you may see a different version of each test that uses the absolute value of the terms. These absolute value versions can be used for series made up of both positive and negative terms. Don't sweat this; the different versions amount to the same thing.)

The ratio test

The ratio test looks at the ratio of a term of a series to the immediately preceding term. If, in the limit, this ratio is less than 1, the series converges; if it's more than 1 (this includes infinity), the series diverges; and if it equals 1, the test is inconclusive.

When to use the ratio test. The ratio test works especially well with series involving *factorials* like *n*! or where *n* is in the power like 3^n.

Definition of the *factorial* symbol. The factorial symbol, !, tells you to multiply like this: $6! = 6 \cdot 5 \cdot 4 \cdot 3 \cdot 2 \cdot 1$. And notice how things cancel when you have factorials in the numerator and denominator of a fraction:

$\frac{6!}{5!} = \frac{6 \cdot 5 \cdot 4 \cdot 3 \cdot 2 \cdot 1}{5 \cdot 4 \cdot 3 \cdot 2 \cdot 1} = 6$ and $\frac{5!}{6!} = \frac{5 \cdot 4 \cdot 3 \cdot 2 \cdot 1}{6 \cdot 5 \cdot 4 \cdot 3 \cdot 2 \cdot 1} = \frac{1}{6}$. In both cases, everything

cancels but the 6. In the same way, $\frac{(n+1)!}{n!} = n+1$ and $\frac{n!}{(n+1)!} = \frac{1}{n+1}$; every-

thing cancels but the $(n+1)$. Lastly, it seems weird, but $0! = 1$ — just take my word for it.

Try this one: Does $\sum\limits_{n=0}^{\infty} \dfrac{3^n}{n!}$ converge or diverge? Here's what you do. You look at the limit of the ratio of the $(n+1)$st term to the nth term:

$$\lim_{n\to\infty} \frac{\dfrac{3^{n+1}}{(n+1)!}}{\dfrac{3^n}{n!}}$$

$$= \lim_{n\to\infty} \frac{3^{n+1} \cdot n!}{(n+1)! \cdot 3^n}$$

$$= \lim_{n\to\infty} \frac{3}{n+1}$$

$$= \frac{3}{\infty+1}$$

$$= 0$$

Because this limit is less than 1, $\sum\limits_{n=0}^{\infty} \dfrac{3^n}{n!}$ converges.

Here's another series: $\sum\limits_{n=1}^{\infty} \dfrac{n^n}{n!}$. What's your guess — does it converge or diverge?

Look at the limit of the $(n+1)$st term over the nth term:

$$\lim_{n\to\infty} \frac{\dfrac{(n+1)^{n+1}}{(n+1)!}}{\dfrac{n^n}{n!}}$$

$$= \lim_{n\to\infty} \frac{(n+1)^{n+1} \cdot n!}{(n+1)! \cdot n^n}$$

$$= \lim_{n\to\infty} \frac{(n+1)^{n+1}}{(n+1) \cdot n^n}$$

$$= \lim_{n\to\infty} \frac{(n+1)^n}{n^n}$$

$$= \lim_{n\to\infty} \left(\frac{n+1}{n}\right)^n$$

$$= \lim_{n\to\infty} \left(1+\frac{1}{n}\right)^n$$

$$= e \qquad (\lim_{n\to\infty} \left(1+\frac{1}{n}\right)^n = e \text{ is one of the limits you should memorize, as discussed in Chapter 8.)}$$

$$\approx 2.718$$

Because the limit is greater than 1, $\sum\limits_{n=1}^{\infty} \dfrac{n^n}{n!}$ diverges.

The root test

Like the ratio test, the root test looks at a limit. This time you investigate the limit of the nth root of the nth term of your series. The result tells you the same thing as the results of the ratio test: If the limit is less than 1, the series converges; if it's more than 1 (including infinity), the series diverges; and if the limit equals 1, you learn nothing.

The root test is a good one to try if the series involves nth powers.

Try this one: Does $\sum\limits_{n=1}^{\infty} \dfrac{e^{2n}}{n^n}$ converge or diverge? Here's what you do:

$$\lim_{n\to\infty} \sqrt[n]{\frac{e^{2n}}{n^n}}$$
$$=\lim_{n\to\infty} \frac{e^{2n/n}}{n^{n/n}}$$
$$=\lim_{n\to\infty} \frac{e^2}{n}$$
$$=\frac{e^2}{\infty}$$
$$=0$$

Because the limit is less than 1, the series converges. By the way, you can also do this series with the ratio test, but it's harder — take my word for it.

Making a good guess about convergence/divergence: Sometimes it's useful to make an educated guess about the convergence or divergence of a series before you launch into one or more of the convergence/divergence tests. Here's a tip that helps with some series. The following expressions are listed from "smallest" to "largest": $n^{10}, 10^n, n!, n^n$. (The 10 is an arbitrary number; the size of the number doesn't affect this ordering.) A series with a "smaller" expression over a "larger" one converges, for example, $\sum\limits_{n=1}^{\infty} \dfrac{n^{50}}{n!}$ or $\sum\limits_{n=1}^{\infty} \dfrac{n!}{n^n}$; and a series with a "larger" expression over a "smaller" one diverges, for instance, $\sum\limits_{n=1}^{\infty} \dfrac{n^n}{100^n}$ or $\sum\limits_{n=1}^{\infty} \dfrac{25^n}{n^{100}}$.

Alternating Series

In the previous sections, you've been looking at series of *positive* terms. Now you look at *alternating series* — series where the terms alternate between positive and negative — like this:

$$1-\frac{1}{2}+\frac{1}{4}-\frac{1}{8}+\frac{1}{16}-\frac{1}{32}+\frac{1}{64}-\cdots$$

Finding absolute versus conditional convergence

Many divergent series of positive terms converge if you change the signs of their terms so they alternate between positive and negative. For example, you know that the harmonic series diverges:

$$1 + \frac{1}{2} + \frac{1}{3} + \frac{1}{4} + \frac{1}{5} + \frac{1}{6} + \dots$$

But, if you change every other sign to negative, you obtain the *alternating harmonic series,* which *converges*:

$$1 - \frac{1}{2} + \frac{1}{3} - \frac{1}{4} + \frac{1}{5} - \frac{1}{6} + \dots$$

By the way, although I'm not going to show you how to compute it, this series converges to ln 2, which equals about 0.6931.

Definition of *conditional convergence:* An alternating series is said to be conditionally convergent if it's convergent as it is but would become divergent if all its terms were made positive.

Definition of *absolute convergence:* An alternating series is said to be absolutely convergent if it would be convergent even if all its terms were made positive. And any such absolutely convergent alternating series is also automatically convergent as it is.

Here's an example. Determine the convergence or divergence of the following alternating series:

$$\sum_{n=0}^{\infty} (-1)^n \frac{1}{2^n} = 1 - \frac{1}{2} + \frac{1}{4} - \frac{1}{8} + \frac{1}{16} - \dots$$

If all these terms were positive, you'd have the familiar geometric series,

$$\sum_{n=0}^{\infty} \frac{1}{2^n} = 1 + \frac{1}{2} + \frac{1}{4} + \frac{1}{8} + \frac{1}{16} + \dots$$

which, by the geometric series rule, converges to 2. Because the positive series converges, the alternating series must also converge (though to a different result — see the following) and you say that the alternating series is *absolutely convergent.*

The fact that absolute convergence implies ordinary convergence is just common sense if you think about it. The previous geometric series of positive terms converges to 2. If you made all the terms negative, it would sum to −2, right? So, if some of the terms are positive and some negative, the series must converge to something between −2 and 2.

Did you notice that the above alternating series is a geometric series *as it is* with $r = -\frac{1}{2}$? (Recall that the geometric series rule works for alternating series as well as for positive series.) The rule gives its sum: $\dfrac{a}{1-r} = \dfrac{1}{1-\left(-\frac{1}{2}\right)} = \dfrac{2}{3}$.

The alternating series test

Alternating series test: An alternating series converges if two conditions are met:

1. Its nth term converges to zero.

2. Its terms are non-increasing — in other words, each term is less than or equal to its predecessor (ignoring the minus signs).

   (Note that you are free to ignore any number of initial terms when checking whether condition 2 is satisfied.)

Using this simple test, you can easily show many alternating series to be convergent. The terms just have to converge to zero and get smaller and smaller (they rarely stay the same). The alternating harmonic series converges by this test:

$$\sum_{n=1}^{\infty} (-1)^{n+1} \frac{1}{n} = 1 - \frac{1}{2} + \frac{1}{3} - \frac{1}{4} + \frac{1}{5} - \frac{1}{6} + \dots$$

As do the following two series:

$$\sum_{n=1}^{\infty} (-1)^{n+1} \frac{1}{\sqrt{n}} = 1 - \frac{1}{\sqrt{2}} + \frac{1}{\sqrt{3}} - \frac{1}{\sqrt{4}} + \frac{1}{\sqrt{5}} - \frac{1}{\sqrt{6}} + \dots$$

$$\sum_{n=1}^{\infty} (-1)^{n+1} \frac{1}{n^2} = 1 - \frac{1}{2^2} + \frac{1}{3^2} - \frac{1}{4^2} + \frac{1}{5^2} - \frac{1}{6^2} + \dots$$

The alternating series test can't tell you whether a series is absolutely or conditionally convergent. The alternating series test can only tell you whether an alternating series itself converges. The test says nothing about the corresponding positive-term series. In other words, the test cannot tell you whether a series is absolutely convergent or conditionally convergent. To answer that, you must investigate the positive series with a different test.

Now try a few problems. Determine the convergence or divergence of the following series. If convergent, determine whether the convergence is conditional or absolute.

$$\sum_{n=3}^{\infty} (-1)^{n+1} \frac{\ln n}{n}$$

1. Check that the nth term converges to zero.

$$\lim_{n \to \infty} \frac{\ln n}{n}$$

$$= \lim_{n \to \infty} \frac{\frac{1}{n}}{1} \quad \text{(by L'Hôpital's rule)}$$

$$= 0$$

Consider the nth term. Always check the nth term first because if it doesn't converge to zero, you're done — the alternating series *and* the positive series will both diverge. Note that the nth term test of *divergence* (see the section on the nth term test) applies to alternating series as well as positive series.

2. Check that the terms decrease or stay the same (ignoring the minus signs).

To show that $\frac{\ln n}{n}$ decreases, take the derivative of the function $f(x) = \frac{\ln x}{x}$. Remember differentiation? I know it's been a while.

$$f'(x) = \frac{\frac{1}{x} \cdot x - \ln x}{x^2} \quad \text{(quotient rule)}$$

$$= \frac{1 - \ln x}{x^2}$$

This is negative for all $x \geq 3$ (because the natural log of anything 3 or greater is more than 1 and x^2, of course, is always positive), so the derivative and thus the slope of the function are negative, and therefore the function is decreasing. Finally, because the function is decreasing, the terms of the series are also decreasing (when $n \geq 3$). That does it:

$$\sum_{n=3}^{\infty} (-1)^{n+1} \frac{\ln n}{n} \text{ converges by the alternating series test.}$$

3. Determine the type of convergence.

You can see that for $n \geq 3$ the positive series, $\frac{\ln n}{n}$, is greater than the divergent harmonic series, $\frac{1}{n}$, so the positive series diverges by the direct comparison test. Thus, the alternating series is *conditionally* convergent.

I can't think of a good title for this warning. If the alternating series fails to satisfy the second requirement of the alternating series test, it does *not* follow that your series diverges, only that this test fails to show convergence.

You're getting so good at this, so how about another problem? Test the convergence of $\sum\limits_{n=4}^{\infty} (-1)^n \dfrac{\ln n}{n^3}$. Because the *positive* series $\dfrac{\ln n}{n^3}$ resembles the convergent *p*-series, $\dfrac{1}{n^3}$, you guess that it converges.

You might want to consider the positive series first. If you think you can show that the *positive* series converges or diverges, you may want to try that before using the alternating series test, because . . .

- ✔ You may have to do this later anyway to determine the type of convergence, and

- ✔ If you can show that the positive series *converges*, you're done in one step, and you've shown that the alternating series is *absolutely* convergent.

So try to show the convergence of the positive series $\sum\limits_{n=4}^{\infty} \dfrac{\ln n}{n^3}$. The limit comparison test seems appropriate here, and $\sum\limits_{n=4}^{\infty} \dfrac{1}{n^3}$ is the natural choice for the benchmark series, but with that benchmark, the test fails — try it. When this happens, you can sometimes get home by trying a larger convergent series. So try the limit comparison test with the convergent *p*-series, $\sum\limits_{n=4}^{\infty} \dfrac{1}{n^2}$:

$$\lim_{n \to \infty} \frac{\dfrac{\ln n}{n^3}}{\dfrac{1}{n^2}}$$

$$= \lim_{n \to \infty} \frac{\ln n}{n}$$

$$= 0 \qquad \text{(We did this in the previous problem with L'Hôpital's rule.)}$$

Because this limit is zero, the positive series $\sum\limits_{n=4}^{\infty} \dfrac{\ln n}{n^3}$ converges (see the section "The limit comparison test"); and because the positive series converges, so does the given alternating series. Thus, $\sum\limits_{n=4}^{\infty} (-1)^n \dfrac{\ln n}{n^3}$ converges *absolutely*.

One last problem and I'll let you go. Test the convergence of $\sum\limits_{n=1}^{\infty} (-1)^{n+1} \dfrac{n}{n+1} = \dfrac{1}{2} - \dfrac{2}{3} + \dfrac{3}{4} - \dfrac{4}{5} + \dfrac{5}{6} - \dots$. This is an easy one.

The nth term of this series (ignoring the minus signs) converges to 1 (it's a L'Hôpital's rule no-brainer), so you're done. Because the nth term does not converge to zero, the series diverges by the nth term test.

Keeping All the Tests Straight

You now probably feel like you know — have a vague recollection of? — a gazillion convergence/divergence tests and are wondering how to keep track of all of them. Actually, I've given you only ten tests in all — that's a nice, easy-to-remember round number. Here's how you can keep the tests straight.

First are the three series with names: the geometric series, p-series, and telescoping series. A geometric series converges if $0 < |r| < 1$. A p-series converges if $p > 1$. A telescoping series converges if the second "half term" converges to a finite number.

Next are the three comparison tests: the direct comparison, limit comparison, and integral comparison tests. All three compare a new series to a known benchmark series. If the benchmark series converges, so does the series you're investigating; if the benchmark diverges, so does your new series.

And then you have the two "R" tests: the ratio test and the root test. Both analyze just the series in question instead of comparing it to a benchmark series. Both involve taking a limit, and the results of both are interpreted the same way. If the limit is less than 1, the series converges; if the limit is greater than 1, the series diverges; and if the limit equals 1, the test is inconclusive.

Finally, you have two tests that form bookends for the other eight — the nth term test of divergence and the alternating series test. These two form a coherent pair. You can remember them as the nth term test of divergence and the nth term test of convergence. The alternating series test involves more than just testing the nth term, but this is a good memory aid.

Well, there you have it: Calculus, schmalculus.

Part VI
The Part of Tens

the
part of
tens

For an extra Part of Tens chapter on cool calculus tips, head on over to
www.dummies.com/extras/calculus.

In this part . . .

✔ Ten simple things you absolutely must know

✔ Ten common mistakes you positively must avoid

✔ Ten things to try as a last resort

Chapter 20

Ten Things to Remember

• •

*T*his chapter contains ten things you should definitely remember. That's not too much to ask, is it? If your mind is already crammed to capacity, you can make some room by first reading Chapter 21, "Ten Things to Forget."

Your Sunglasses

If you're going to have to study calculus, you might as well look good.

If you wear sunglasses *and* a pocket protector, it'll ruin the effect.

$a^2 - b^2 = (a - b)(a + b)$

This factor pattern is quasi-ubiquitous and somewhat omnipresent; it's used in a plethora of problems and forgetting it will cause a myriad of mistakes. In short, it's huge. Don't forget it.

$\frac{0}{5} = 0$, but $\frac{5}{0}$ Is Undefined

$\frac{8}{2} = 4$, and so 4 times 2 is 8. If $\frac{5}{0}$ had an answer, that answer times zero would have to equal 5. But that's impossible, making $\frac{5}{0}$ undefined.

Anything^0 = 1

The only exception is 0^0, which is undefined. The rule holds for *everything* else, including negatives and fractions. This may seem a bit weird, but it's true.

SohCahToa

No, this isn't a famous Indian chief, just a mnemonic for remembering your three basic trig functions:

$$\sin\theta = \frac{O}{H} \qquad \cos\theta = \frac{A}{H} \qquad \tan\theta = \frac{O}{A}$$

Trig Values for 30, 45, and 60 Degrees

$$\sin 30° = \frac{1}{2} \qquad \sin 45° = \frac{\sqrt{2}}{2} \qquad \sin 60° = \frac{\sqrt{3}}{2}$$

$$\cos 30° = \frac{\sqrt{3}}{2} \qquad \cos 45° = \frac{\sqrt{2}}{2} \qquad \cos 60° = \frac{1}{2}$$

$$\tan 30° = \frac{\sqrt{3}}{3} \qquad \tan 45° = 1 \qquad \tan 60° = \sqrt{3}$$

$sin^2\theta + cos^2\theta = 1$

This identity holds true for *any* angle. Divide both sides of this equation by $\sin^2\theta$ to get $1 + \cot^2\theta = \csc^2\theta$; divide by $\cos^2\theta$ to get $\tan^2\theta + 1 = \sec^2\theta$.

The Product Rule

$\frac{d}{dx}(uv) = u'v + uv'$. Piece o' cake.

The Quotient Rule

$\frac{d}{dx}\left(\frac{u}{v}\right) = \frac{u'v - uv'}{v^2}$: In contrast to the product rule, many students forget the quotient rule. But you won't if you just remember to begin the answer with the derivative of the top of your fraction, *u*. This is easy to remember because it's the most natural way to begin. The rest falls into place.

Where You Put Your Keys

No one can predict what score you'll get on your next calculus exam — unless, that is, you don't show up.

Chapter 21

Ten Things to Forget

• •

*T*his is without question the easiest chapter in the book. There's nothing to study, nothing to comprehend, nothing to learn. Just kick back, crank up the music, and *forget* this stuff.

$(a+b)^2 = a^2 + b^2$ — *Wrong!*

Don't confuse this with $(ab)^2 = a^2 b^2$, which is *right*. $(a+b)^2$ equals $a^2 + 2ab + b^2$.

$\sqrt{a^2 + b^2} = a + b$ — *Wrong!*

Don't confuse this with $\sqrt{a^2 b^2} = ab$, which is *right*. $\sqrt{a^2 + b^2}$ can't be simplified.

$Slope = \dfrac{x_2 - x_1}{y_2 - y_1}$ — *Wrong!*

This is upside down. Slope equals $\frac{y_2 - y_1}{x_2 - x_1}$.

$\dfrac{3a+b}{3a+c} = \dfrac{b}{c}$ — *Wrong!*

You can't cancel the $3as$ because it's not a *factor* of the numerator and the denominator. With $\frac{3ab}{3ac} = \frac{b}{c}$, however, you *can* cancel the $3as$.

$\dfrac{d}{dx}\pi^3 = 3\pi^2$ — *Wrong!*

Pi (π) is a number, not a variable, so π^3 is also just a number, and the derivative of any number is zero. Thus, $\frac{d}{dx}\pi^3 = 0$.

If k Is a Constant, $\frac{d}{dx}kx = k'x + kx'$ — Wrong!

You don't use the product rule here. Constants work like numbers, not variables, so $\frac{d}{dx}kx$ works just like $\frac{d}{dx}3x$, which equals 3. Thus, $\frac{d}{dx}kx = k$. (Extra credit: Do you see why, in fact, the above is not technically wrong?)

The Quotient Rule Is $\frac{d}{dx}\left(\frac{u}{v}\right) = \frac{v'u - vu'}{v^2}$ — Wrong!

See the second from the last point in Chapter 20, "Ten Things to Remember."

$\int x^2\,dx = \frac{1}{3}x^3$ — Wrong!

Do you C why this is wrong?

$\int (\sin x)\,dx = \cos x + C$ — Wrong!

The derivative of cosine is *negative* sine, so the derivative of *negative* cosine is sine, and thus $\int (\sin x)\,dx = -\cos x + C$.

Green's Theorem

$$\int_c (Mdx + Ndy) = \iint_R \left(\frac{\partial N}{\partial x} - \frac{\partial M}{\partial y}\right) dA$$

This one's *right,* but forget trying to remember it.

Chapter 22

Ten Things You Can't Get Away With

• •

*T*he original title of this chapter was "Ten Things You Can Get Away With If Your Calculus Teacher Was Born Yesterday," but the legal department was afraid someone would actually try some of these stunts, get caught, and then file a lawsuit. So they changed the title to the boring one you have now.

Give Two Answers on Exam Questions

If you can't make up your mind about which of two answers is correct, put them both down with both of them sort of circled and both sort of crossed out. If one of your two answers is correct, your teacher will give you the benefit of the doubt.

Write Illegibly on Exams

Get an answer on your calculator and then scribble your "work" so sloppily that your teacher can't read it. Because you got the correct answer, he'll assume that you knew what you were doing and give you full credit.

Don't Show Your Work on Exams

Get an answer on your calculator and write the following next to the problem, "Easy problem — did work in my head." Your teacher will take your word for it.

Don't Do All of the Exam Problems

If an exam is, say, four pages long and stapled together, find the page with the worst-looking problems on it, remove the staple, put the bad page in your pocket, and carefully replace the staple. Your teacher will assume that the page was omitted at the copy center. When you later complete the "missing" part of the test perfectly, your teacher will suspect nothing.

Blame Your Study Partner for Low Grade

Tell your teacher that the person you studied with explained everything wrong, so it's not your fault. Your teacher will let you retake the exam.

Tell Your Teacher You Need an "A" in Calculus to Impress Your Significant Other

Your teacher, being a romantic at heart — and remembering his days as an undergraduate when he aced calculus and then became a babe magnet — will give you the "A."

Claim Early-Morning Exams Are Unfair Because You're Not a "Morning Person"

Explain that your biological clock is out of sync with your school's old-school, early-to-bed-early-to-rise Protestant ethic. Your teacher will let you take your exams in the afternoon and trust you to not talk with friends who take the morning exams.

Protest the Whole Idea of Grades

Make a political stink about teachers who have the nerve to presume that they have the right to give you a grade. Who are they to be evaluating *you?* Claim to be a conscientious objector when it comes to grades. Argue that giving grades reflects an unfair talent and intelligence bias — that the whole system is classist and IQist. Your teacher will be impressed with the depth and sincerity of your philosophical convictions and will let you take all exams pass/fail.

Pull the Fire Alarm During an Exam

This one's a bit juvenile — in contrast, of course, to the preceding tips.

Use This Book as an Excuse

If you get caught trying these stunts, tell your teacher that you thought it was okay because you read it in a book. Your teacher will let you off the hook.

Index

• E •

• **N** •

• **O** •

• **P** •

About the Author

A graduate of Brown University and the University of Wisconsin Law School, **Mark Ryan** has been teaching math since 1989. He runs The Math Center in Winnetka, Illinois (`www.themathcenter.com`), where he teaches junior high and high school math courses including an introduction to calculus. He also does extensive one-to-one tutoring for all levels of mathematics and for standardized test preparation. In high school, he twice scored a perfect 800 on the math portion of the SAT, and he not only knows mathematics, he has a gift for explaining it in plain English. He practiced law for four years before deciding he should do something he enjoys and use his natural talent for mathematics. Ryan is a member of the Authors Guild and the National Council of Teachers of Mathematics.

Calculus For Dummies, 2nd Edition is Mark Ryan's eighth book. His first book, *Everyday Math for Everyday Life* (Grand Central Publishing), was published in 2002. For Wiley, *Calculus For Dummies,* 1st Edition was published in 2003, *Calculus Workbook For Dummies* in 2005, *Geometry Workbook For Dummies* in 2007, *Geometry For Dummies,* 2nd Edition in 2008, *Calculus Essentials For Dummies* in 2010, and *Geometry Essentials For Dummies* in 2011. Ryan's math books have sold over a half of a million copies.

Dedication

To my current and former math students. Through teaching them, they taught me.

Author's Acknowledgments

I'm very grateful to my agent, Sheree Bykofsky, and her staff for all their efforts which ultimately resulted in my writing *Calculus For Dummies* for Wiley Publishing. A special thanks to my brother-in-law, Steve Mardiks, and my friends Abby Lombardi, Ted Lowitz, and Barry Sullivan for their valuable advice, editing, and support. Josh Dillon and Jason Molitierno did a great job checking the calculus content of the book as well as the clarity of exposition.

Everyone at Wiley has been great to work with. Acquisitions Editor Kathy Cox has a refreshing, non-bottom-line desire to serve the best interests of the reader. Acquisitions Editor Lindsay LeFevere is a delight to work with. She's professional, smart, down-to-earth, and has a great sense of humor. She always has an empathetic and quick understanding of my concerns as an author. Editors Tim Gallan, Laura Peterson, and Corbin Collins have just the right mix of patience and a stick-to-the-deadline approach. These talented editors understand the forest, the trees, when to edit, and when not to edit. A special thanks to Corbin Collins, who — considering the inevitable challenges that arise in producing a book — has a talent for making the process fun. Krista Fanning did a great job with the book's thousands of equations and mathematical symbols. And the layout and graphics team did an excellent job with the book's complex math. This book is a testament to the high standards of Wiley Publishing.

Publisher's Acknowledgments

Acquisitions Editor: Lindsay LeFevere

Editor: Corbin Collins

Technical Editor: Jason Molitierno, PhD

Project Coordinator: Melissa Cossell

Illustrations: John Wiley & Sons, Inc.

Cover Image: ©iStockphoto.com/Henvry